AIChE Equipment Testing Procedure
Trayed and Packed Columns

AIChE Equipment Testing Procedure
Trayed and Packed Columns

A Guide to Performance Evaluation
Third Edition

Equipment Testing Procedures Committee of the
American Institute of Chemical Engineers

Cover and book design: Lois Anne DeLong

A Joint Publication of the Center for Chemical Process Safety of the American Institute of Chemical Engineers and John Wiley & Sons, Inc.

Published by John Wiley & Sons, Inc., Hoboken, New Jersey

Published simultaneously in Canada

For general information on our other products and services or for technical support, please contact our Customer Care Department within the United States at (800) 762-2974, outside the United States at (317) 572-3993 or fax (317) 572-4002.

Wiley also publishes its books in a variety of electronic formats. Some content that appears in print may not be available in electronic formats. For more information about Wiley products, visit our web site at www.wiley.com.

Library of Congress Cataloging-in-Publication Data:

American Institute of Chemical Engineers
    AIChE Equipment Testing Procedure : Trayed and packed columns / a guide to performance evaluation, third edition.
    p. cm.
  Includes index.
ISBN 978-1-118-62771-6 (paper)

Printed in the United States of America

10 9 8 7 6 5 4 3 2 1

AMERICAN INSTITUTE OF CHEMICAL ENGINEERS EQUIPMENT TESTING PROCEDURES COMMITTEE

Chair: **Becky Starkweather, P.E.**
Scientex L.C.

Vice Chair: **Prashant D. Agrawal, P.E.**
Consultant

Past Chair: **James Fisher, P.E.**
Amgen, Inc.

TRAYED & PACKED COLUMNS PROCEDURE REVISION SUBCOMMITTEE

Chair: **Zhanping Xu**
UOP LLC, A Honeywell Company

Co-Chair: **Daniel R. Summers**
Sulzer Chemtech USA, Inc

General Committee Liaison: **Prashant D. Agrawal, P.E.**
Consultant

WORKING COMMITTEE

Tony Cai
Fractionation Research, Inc.

Bruce Holden
Dow Chemical

Ron Olsson
Celanese Corp.

Lowell Pless
Tracerco

Michael Schultes
Raschig GmbH

Frank Seibert
University of Texas

Simon X. Xu
Technip

CONTRIBUTING REVIEWERS

Greg Cantley
Marathon Oil

Brad Fleming
Raschig USA

Joe Flowers
DuPont

Dennis Maloney
CB&I Lummus

Todd Marut
ExxonMobil

Paul Morehead
Koch-Glitsch

Zarko Olujic
Univ. of Delft, NL

Toshiro Wakabayashi
Toyo Engineering, JP

Tim Zygula
BASF

Company affiliations are shown for information only and do not imply Procedure approval by the company listed.

AIChE Equipment Testing Procedure

Acknowledgement: This procedure draws heavily on the *AIChE Tray Distillation Columns Testing Procedure, 2nd Edition, 1987* and *Packed Columns Testing Procedure, 2nd Edition, 1990*.

Second Edition
　Packed Columns: Officially approved for publication by AIChE Council in 1990
　Tray Distillation Columns: Officially approved for publication by AIChE Council in 1987

First Edition
　Packed Columns: Officially approved for publication by AIChE Council in 1965
　Tray Distillation Columns: Officially approved for publication by AIChE Council in 1962

Members Participating in the Second Edition

Packed Columns

J.W. Drew, L.O. Frescoln, T.L. Holmes, C-L. Hsieh, H.Z. Kister,
G.J. Kunesh, T.P. Ognisty

Tray Columns

M.L. Becker, R.M. Bellinger, O. Frank, R. Harrison, J.L. Humphrey,
E.J. Lemieux, E.J. Noelke, M. Sakata, V.C. Smith

Members Participating in the First Edition

Packed Columns

C.S. Brown, R.S. Eagle, T. Liggett, O.W. Ridout, H.L. Shulman

Tray Distillation Columns

K.L. Birk, G.K. Connolly, L.H. Corn, R.S. Eagle, J.A. Gerster, E.W. Grohse, J.E. Jubin,
E.T. Merrill, M.P. Nelson, R.F. Romell, E.H. Ten Eyek, T.J. Walsh

The rewrite committee would like to acknowledge Henry Kister for his contribution to a major portion of this work. Even though he was not officially on the re-write committee, his wording on test procedure from the second edition of AIChE's Packed Columns Equipment Testing Procedure were retained nearly word for word, as can be exemplified in pages 388-419 of his 1990 book Distillation Operation. When combining both the Tray Distillation Columns and Packed Columns testing procedures, the re-write committee purposefully retained the words generated by Henry Kister in Section 500.0 from "Packed Columns" because of its extensive content and clarity.

CONTENTS

List of Tables

List of Figures

100.0 PURPOSE AND SCOPE

101.0 Purpose

This testing procedure offers methods of conducting and interpreting performance tests on trayed and packed distillation columns. Such tests are intended to accumulate reliable data in one or more of the following areas of interest: mass transfer efficiency, capacity, energy consumption and pressure drop. It is intended to be used as a guideline for a column performance test and not as a substitute for a vendor's acceptance test.

Possible uses of such data include:

- Troubleshooting performance problems

- Identifying capacity bottlenecks

- Determining if column performance meets vendor guarantees ("acceptance test")

- Developing basic data for new designs

- Developing correlations

- Determining the operating range of a column

- Defining optimum operating conditions

- Calibrating computer simulations for use in optimizing, debottlenecking and design studies.

102.0 Scope

Rather than compulsory directions, this book offers a collection of techniques presented to guide the user, and emphasis is placed on principles, rather than on specific steps. It applies to columns that operate either at steady state or at total reflux. It does not apply to batch columns in which compositions are changing with time unless they are operated at total reflux or with distillate returned to the still pot during the test.

The procedure applies to both trayed and packed columns of any type. The tests determine the composite performance of the trays, packing, and any associated distributors and other auxiliary internals inside the column. It is important to realize that capacity may be restricted by these auxiliaries, particularly for packed towers.

This procedure does not apply to external testing of distributors or other internals.

200.0 DEFINITION AND DESCRIPTION OF TERMS

201.0 Flow Quantities (Refer to Figure 1)

201.1 *Feed* is the material to be separated, including multiple feed streams.

201.2 *Bottoms* describes the high-boiling product leaving the bottom of the column (or the reboiler).

201.3 *Distillate* is the product distilled overhead. It may leave the distillation system as a vapor, liquid, or a combination of both.

201.4 *Side-stream Product* is product withdrawn from an intermediate section of the column.

201.5 *Overhead Vapor* designates the vapor from the top of the column and includes material to be condensed for reflux. It is the combined distillate and external reflux.

201.6 *Reflux* is used to designate the quantity of liquid returned to the column.

> **201.6.1** *External (Overhead) Reflux* is the quantity of liquid returned to the top of the column. External reflux may be subcooled, which can result in increased internal reflux.
>
> **201.6.2** *Internal Reflux* is the calculated quantity of liquid leaving the top theoretical stage inside the tower. The internal reflux is different from the external reflux in that it is in thermal equilibrium with the top theoretical stage inside the tower.
>
> **201.6.3** *Pumparound* is the quantity of liquid withdrawn from, and returned to, the column after being cooled. A pumparound can be subcooled, and then returned to the tower at a location other than the top. A pumparound is sometimes called *Circulating Reflux*.
>
> **201.6.4** *Reflux Ratio* is the ratio of the external reflux flow to the distillate. Some applications may use the ratio of external reflux flow to feed to represent the reflux ratio.

201.7 *Throughput* refers to the combined liquid and vapor traffic passing through a cross section of the column.

> **201.7.1** *Internal Liquid* is the calculated quantity of liquid flowing from point to point in the column.
>
> **201.7.2** *Internal Vapor* is the calculated quantity of vapor passing from point to point in the column.
>
> **201.7.3** *Entrainment* is the liquid carried upward by the vapor stream from one point to another.
>
> **201.7.4** *Weeping* is the liquid that flows downward through the deck openings in trayed towers.

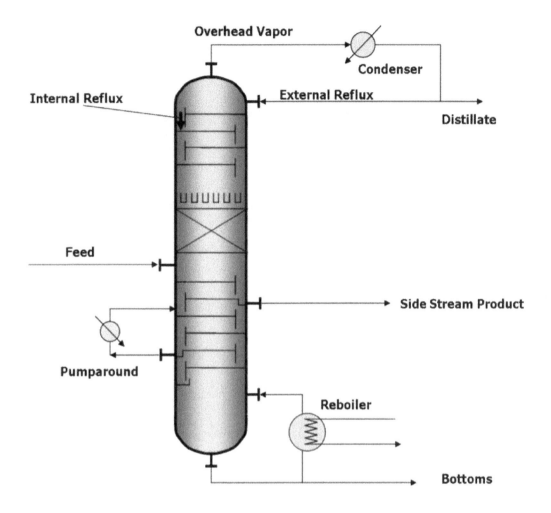

Figure 1. Graphic Description of Terms

202.0 Key Components

A *key component* is a component of interest in a column. In a multi-component mixture, separation is based upon the lower boiling "light key" and the higher boiling "heavy key." At times components with boiling temperatures between those of the light and heavy key may also be present. Such a component may be called a "middle boiler," an "intermediate," or a "distributed key."

203.0 Mass Transfer Efficiency

The efficiencies used in describing the performance of fractionation columns are briefly described below [1].

203.1 *Theoretical Trays or Plates or Stages* are stages on which the vapor and liquid streams leaving the stage have reached thermodynamic equilibrium.

203.2 *Overall Column Efficiency* refers to the performance of the column as a whole. For trayed columns, this efficiency is the ratio of the number of theoretical trays or stages that would be required for the observed separation to the number of actual trays in the column. This number is dimensionless and is usually expressed as a percentage.

For packed columns, the *HETP (Height Equivalent to a Theoretical Plate)* for the column as a whole is the ratio of the overall height of packing in the column to the number of theoretical stages in the column. This number has units of length per theoretical plate (or stage).

For calculation of overall efficiency, the stages are those developed by the trays and/or packing. Any stages attributed to a reboiler or a partial condenser should not be included in this calculation. Overall efficiency is useful in comparing one test with another, or in comparing the test with design. The overall efficiency of sections of a column may also be of interest. For example, the overall efficiency below the feed point may be different from the overall efficiency above the feed.

203.3 *Apparent Murphree Tray Efficiency* can be measured by taking samples around a single tray. It is the actual change in composition accomplished by the tray divided by the change that would occur on a theoretical tray based on exit liquid composition. It accounts for the effects of entrainment, weeping, liquid mixing, maldistribution, and other factors.

203.4 *Ideal Murphree Tray Efficiency* is the performance of a single tray, exclusive of the deleterious effects of entrainment, weeping, and liquid backmixing. The ideal Murphree tray efficiency can be predicted from the Murphree point efficiency.

203.5 *Murphree Point Efficiency* is the Murphree efficiency at a single point on the tray.

203.6 *HETP (Height Equivalent to a Theoretical Plate)* is the height of packing required to perform as one theoretical plate or stage of separation.

203.7 *HTU (Height of a Transfer Unit)* is the height of packing required to obtain one transfer unit and is a measure of the mass transfer efficiency. As such, it incorporates mass transfer coefficients. The more efficient the mass transfer (i.e., larger mass transfer coefficient and/or larger vapor-liquid contacting area), the smaller the value of HTU. This value can be estimated from empirical correlations, or measured directly from pilot plant tests (See Section 605.2.2).

203.8 *NTU (Number of Transfer Units)* is a measure of the difficulty of the separation. A single transfer unit gives the change of composition of one of the phases equal to the average driving force producing the change. The NTU is similar to the number of theoretical stages or plates required for trayed columns. A larger number of transfer units is required for a higher purity product (See Section 605.2.2).

204.0 Operating Lines

Operating Lines are the material-balance lines on a McCabe-Thiele type of diagram for a binary system [2]. The use of operating lines has been extended to multi-component systems by distributing the non-key components to the key binary components [3].

205.0 Pinch

Pinch is a term that describes a local condition within the column under which no appreciable change in composition of the liquid or vapor from stage to stage occurs due to a lack of mass transfer driving force, and not to column malfunctioning, flooding, high entrainment, or a dry tray or packing. For a binary system, a pinch is graphically depicted when an operating line is approaching or intersects the equilibrium curve on a McCabe-Thiele diagram.

206.0 Maximum Throughput

206.1 *Maximum Hydraulic Throughput* is the highest loading at which a column can operate without flooding. Since the loading is affected by both liquid and vapor rates, many combinations of those rates may define a maximum hydraulic throughput curve. In practice, this maximum may be limited by the trays, or the packed bed, or one of the internals.

> **206.1.1** *Flooding* describes the condition of the column when the hydraulic capacity is exceeded. At loadings achieving or reaching the flooding point, liquid accumulates uncontrollably and continued operation becomes impossible. The flooding point depends on both vapor and liquid velocities, system properties, and tray or packing geometry [4]. The flooding point can be recognized by the existence of one or more of the situations described in Section 502.1.

206.2 *Maximum Operational Capacity* is the highest loading at which stable operation leading to acceptable overall column efficiency is achieved. This point may occur well below the column's maximum hydraulic capacity. Since loading involves both liquid and vapor, a maximum operational capacity curve can be defined on a plot of liquid vs. vapor rates.

206.3 *Maximum Efficient Capacity* is a term applied most commonly to random packing, where HETP is typically flat with respect to capacity, and then decreases as flood is approached before increasing again. The maximum efficient capacity is the last point at which the HETP of the packing is equivalent to its value in the flat part of the curve.

207.0 Minimum Operating Rate

Minimum Operating Rate is the smallest loading at which a column performance is acceptable. The separation may become unacceptable because of loss of efficiency due to excess weeping of the tray, incomplete wetting of the packing, liquid or vapor maldistribution, or column instability.

208.0 Operating Section

An *Operating Section* of a column is a part to which no feed is added, no product is removed, and no external heat is added or removed. It may consist of one or more beds of packing, or multiple trays.

209.0 Hardware

209.1 Components of a Trayed Column

A trayed column consists of one or more sections of trays, each separated by an inlet and/or outlet stream that can be liquid, vapor or a mixture of both.

> **209.1.1** *Trays* are generally horizontal plates that enable vapor and liquid to contact each other. Trays are composed of decks with (or without) downcomers.
>
> **209.1.2** *Downcomers* are conduits through which liquid passes from one tray to another.
>
> **209.1.3** *Decks* are typically the horizontal active portions of trays and usually contain vapor dispersers (i.e., valves, sieve holes, etc.). Tray decks can be tilted slightly off horizontal.
>
> **209.1.4** *Liquid Distributors* are devices installed at the top of a trayed section by which liquid entering the section is distributed to decks or downcomers in proportion to the column cross-sectional area these components cover. The complexity of the liquid distributors is dependent on the type of trays to which the liquid is distributed.
>
> **209.1.5** *Vapor Distributors* and *Flashing Feed Distributors* are devices for distributing vapor and/or liquid. Vapor distributors are installed below a trayed section.

Besides the above basic components, modern high-capacity trays or devices may contain additional elements, such as vapor-liquid separation devices that increase tray capacity by enhancing vapor/liquid separation between trays. These devices may include enhanced tray periphery vapor/liquid contacting, vapor to liquid momentum transfer devices, and froth promoters.

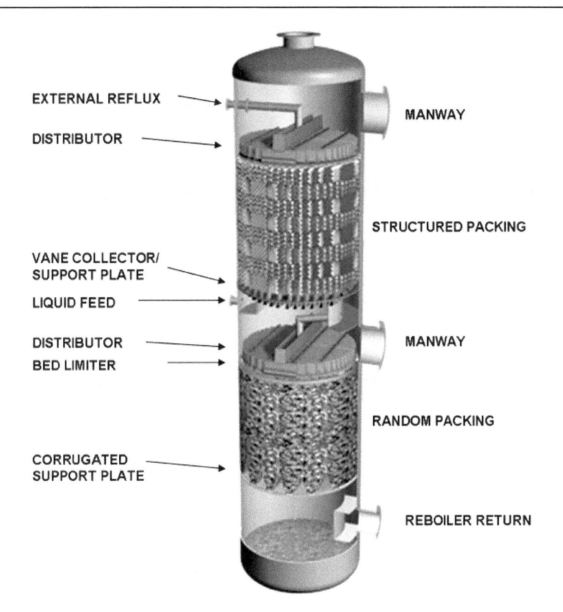

Figure 2. Packed Column Components

209.2 Components of a Packed Column

A packed column consists of one or more sections of packing, each having internals consisting of a liquid distributor, generally framed by a bed limiter or a hold down plate at the top, and a support plate at the bottom (See Figure 2).

> **209.2.1** *Packing* is randomly dumped or structured material inserted inside a column to affect vapor and liquid contact.

209.2.2 *Support Plates* are devices installed at the bottom of a packed bed to support the packing while allowing vapor and liquid to flow freely into and out of the bed.

209.2.3 *Bed Limiters* are devices fixed in place above a packed section to retain packing elements at high vapor flow rates. They are used with metal or plastic packings and may be part of the liquid distributor.

209.2.4 *Hold-down Grids* are similar to bed limiters but are used with ceramic packings. They are sometimes not supported from the column wall but rest directly on the packing, preventing packing fluidization. Hold-down grids may also be used with structured packing to provide uplift resistance and they may be anchored to the vessel wall.

209.2.5 *Liquid Distributors* are devices installed at the top of a packed section by which liquid entering the section is distributed uniformly over the top of the packing.

209.2.6 *Liquid Redistributors* are devices used to improve column performance by collecting and redistributing liquid flowing down from a packed section above.

209.2.7 *Vapor Distributors* are devices installed at the bottom of a packed section by which vapor entering the section is distributed uniformly over the bottom of the packing.

209.2.8 *Flashing Feed Distributors* aid in the disengaging of vapor from liquid, distribute incoming vapor to the underside of a packed bed, and direct the liquid to the liquid distributor for the packed bed below.

300.0 TEST PLANNING

301.0 Preliminary Preparation

The cost of doing plant tests goes far beyond the time and material expended during the actual test run. Careful planning and preparation are essential to maximize the economic and technical benefits of a test. (See Shaw, Sykes and Ormsby [5] and Kister [6] for an in-depth discussion of preliminary test preparation.)

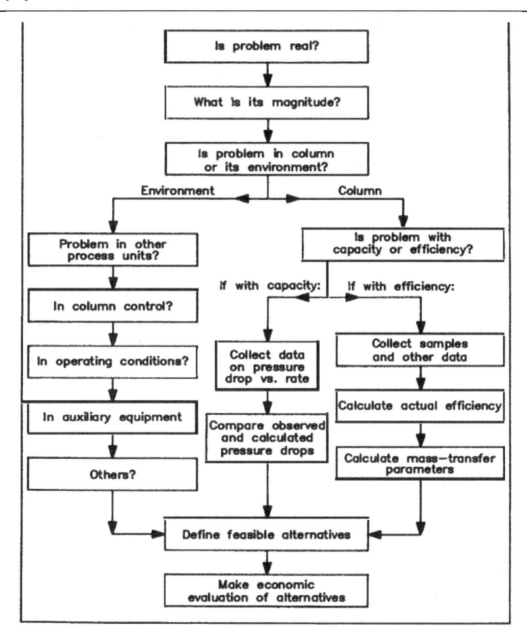

Figure 3. Logic Diagram for Column Troubleshooting

301.1 Safety

Any equipment testing must conform to the latest requirements of applicable safety standards. These include, but are not limited to, plant, industry, local, state, and federal regulations. It is recommended that all testing be conducted under the supervision of personnel fully experienced in plant and equipment operating practices.

301.2 Environmental Considerations

The test procedure must conform to the latest requirements of applicable environmental standards that include plant, industry, local, state, and federal regulations. The same environmental conditions that apply to the equipment in normal operation should also apply during testing.

301.3 Test Objectives

Specific objectives should be stated for the plant test. If the test is proposed for plant troubleshooting, a logic diagram similar to Figure 3 is useful for defining the nature of the problem, and formulating test objectives [7]. This phase of test planning should include techniques for determining whether the objectives are being met while the test is in progress. A test plan and procedure should include detailed test objectives, along with an explanation of exactly how the proper test conditions will be obtained.

301.4 Organizational Resources

The test objectives will help determine which organizational resources should be involved in the test. Plant tests frequently involve people from many disciplines, such as plant maintenance, analytical services, research, design organization representatives, outside service providers and equipment supplier representatives.

301.5 Schedule

Test scheduling should be given the earliest possible consideration. Ideally, the test should scheduled for a time when the plant is expected to run at the desired rates, under as stable conditions as possible. Furthermore, the test should normally be scheduled at a time when unit raw materials are reasonably constant. However, it should be realized that the test may be conducted under extremely unfavorable conditions, and alternate plans should be provided. Feedstock and product inventories should be compatible with potential increased or decreased production rates resulting from the test. The length of the test will depend on its objectives, the number of conditions tested, and the time required to reach steady state for each test condition (See Section 503.0). The test schedule may also be affected by the number of samples to be collected and methods of sample analysis, since laboratory support is frequently a "bottleneck" in operating plants. For columns having different summer and winter operating conditions, the possibility of season-dependent test results should also be considered.

301.6 Review of Historic Operation Data

To understand the current operating characteristics, test planners must obtain heat and material balance information from the column. This information includes, but is not limited to, pressure drop, temperature profile, reflux, feed, product, vent, steam rates, base liquid level, lab analysis, online analyzers, reboiler data and anything else that may provide an insight into the column's current behavior. Normally, operating data can be obtained from electronic data historians, distributed control systems or, in older plants, even in strip charts.

The next step is to reconcile the column data. Usually the material balance, calculated from available instruments, will not close perfectly. For example, reflux and steam may not match and a method must be developed to decide which one is the more accurate by verifying the data through other measurements. If lab analysis or online analyzers are used, the relative accuracy of these instruments for the critical components should be confirmed [8].

302.0 Column Control and Instrumentation

The column control scheme should be well understood by the test participants so that the column's correct response to deliberate or consequential disturbances can be anticipated. Equipment diagrams, which show the locations for temperature and pressure measurement and sample points, should be available. The accuracy of column instrumentation is crucial to a successful test. Critical instruments should be recalibrated before starting the test, and, if necessary, controllers should be retuned for test stability. If direct flow metering is not possible on critical streams, plans should be devised for calculating the flows based on known data. Where flow meters are not installed, other flow measuring services, such as external bolt-on meters, could be considered, and test participants should verify the capability for tight shut-off by manual block valves so that total reflux is, in fact, total. Preparations must be made for piping, instrument, and equipment changes where necessary to obtain the operating conditions, samples, and data required.

303.0 Peripheral Equipment

Potential limitations of peripheral equipment, such as reboilers, condensers, pumps and valves, should be considered, along with effects of the test on downstream equipment or other processing units. This is especially important when testing distillation columns that are thermally integrated with other operations. A good supply of feed at the desired compositions, temperature and pressure must also be available at a correct and steady rate.

304.0 Pre-test Calculations

304.1 Process Simulation

When computer models of the process are available, it is helpful to run a simulation of anticipated test conditions. Flow diagrams showing enthalpy and material balances are instructive for test participants, as such simulations may identify key debottlenecking parameters—like column pressure or reflux ratio—and help to define the ranges of the tested variables. Frequently, results from simulating a sequence of step changes in operating conditions can be shown graphically. These graphs are useful for comparing actual and anticipated results while the test is in progress. Another useful pretest calculation is a hydraulic analysis to determine the expected capacity limitation and pressure drop for all zones in the entire column. The importance of these calculations is discussed in Sections 308.0, 502.1 and 604.0. Pretest calculations will require the accumulation of physical and thermodynamic property data, as summarized in Section 309.0.

304.2 Dry Run

Conducting a "dry run" before the official test is a useful technique. This pretest, followed by rough processing of the data, enables one to spot shortcomings in data collection; detect factors causing poor closure of mass, component and enthalpy balances; and validate the computer simulation (if available) to be used for analyzing test data.

305.0 Types of Tests

305.1 *Performance tests* are conducted to obtain data in the following areas:

- Tray and/or packing efficiency
- Capacity and capacity limitations
- Energy consumption
- Pressure drop

The data collected can be useful for troubleshooting, new designs, validating correlations, and determination of the operating range of a column, as well as the optimum operating conditions.

305.2 An *Acceptance test* is a special type of performance test. Ideally, the conditions under which the acceptance test is run should be those for which the column was designed or guaranteed. The information generated by this test may be needed to check that the compositions and production capacity of the distillate and bottoms products, and the utilities consumption, is as designed or guaranteed.

Those responsible for operating the unit should try to achieve design feed rates and compositions for a reliable acceptance test. Frequently, however, the conditions for the test cannot be the same, in all respects, as those for which the equipment was designed. Here a test can still be made to check the column, and test results can be extrapolated to the design conditions by techniques based on sound engineering principles. Any variations in the test conditions should be discussed by the organizations associated with the test, and should be agreed to before the test is conducted.

306.0 Specific Areas of Interest

In conducting efficiency tests, every possible source of error must be minimized. Testing at total reflux minimizes errors in heat input, feed point composition, and other data required to define operating lines. Simplifying efficiency tests by performing them at total reflux is not without its price, since the results may differ from those at a finite reflux ratio. When both cannot be performed, the choice depends upon the information desired. For determining optimum design conditions, and packing height or number of trays for new designs, finite reflux ratio tests are recommended.

306.1 Packing Efficiencies

The interest in packing efficiency *(HETP, HTU)* may be limited to overall column efficiency or may include efficiencies for several or all sections of a column. HETP or HTU is determined based on key components, and the assumption that vapor-liquid equilibrium data are accurate.

306.2 Tray Efficiencies

The interest in tray efficiency may be limited to overall column efficiency, or may include apparent Murphree tray efficiencies for several or all sections of a column.

> **306.2.1** *Murphree tray efficiencies* are determined based on key components, and the assumption that vapor-liquid equilibrium data are accurate. If the performance characteristics of a particular type of tray are wanted, it is necessary to obtain data to calculate the apparent Murphree tray efficiency. The only reliable method of obtaining such data is to take samples from enough trays in a column to establish a complete concentration gradient. This is seldom practical in a commercial column. Even then, the only

regions where good tray efficiency data may be obtained are those where appreciable changes in the concentrations of the key components take place from tray to tray.

306.3 Overall Column Efficiency

Results from overall column efficiency tests can be useful in determining optimum feed location, optimum reflux ratio, or proper design for new equipment in a similar service.

If possible, feeding to a different location can often result in better separation of the key components. Where multiple feed points have been furnished on the column, tests should be made at various feed locations to determine the optimum.

Pinched conditions at the feed location should be avoided if the test is to measure the true separating ability of the column. A constant temperature zone above or below the feed location may indicate a pinch at this point. This type of pinch can frequently be overcome by increasing the reflux ratio. The amount of pinching that can be tolerated at the top or bottom depends on how accurately the products can be analyzed. Pinch points can also be avoided by relocating the feed, but this is rarely practical in a packed column.

Overall, column efficiency tests are valuable for comparison with efficiencies obtained from other columns operating on the same system at the same relative approach to flooding. They are also useful for comparison with predicted efficiencies, such as from the O'Connell correlation for trays (see Figure 21), to get a sense of how well the column is performing vs. expectations.

306.4 Capacity Limitations

Knowledge of the maximum and minimum capacity of a column is useful when the system is subject to internal load variations, such as from feed and reflux changes, or when future expansion plans or throughput reductions are contemplated.

> **306.4.1** *Maximum Hydraulic Throughput:* At maximum throughput for stable operation, the upper limit of vapor-liquid flow capacity at one point in the column has nearly been reached. Above this rate, the column will begin to flood. It should be realized that the maximum vapor loading depends upon the liquid rate. Furthermore, the vapor and liquid rates usually vary from one section of the column to another. Therefore, several types of capacity tests should be considered, if auxiliaries permit.
>
> a. Determination of the maximum feed rate while reflux and reboil rates are proportionately increased.
>
> b. Determination of the maximum reboil rate with reflux to balance at a constant feed rate.
>
> Note that, in packed columns, distributor and support plates may reach capacity limits before the packing itself. (See Sections 502.0 and 604.0)
>
> Two additional tests may be useful if a feed preheater is in service:
>
> a. Determination of the capacity of the column section above the feed by increasing the preheat, even into the zone of partial feed vaporization— if

13

the feed point can tolerate a two-phase condition—and reflux to balance.

b. Determination of the capacity of the column section below the feed by increasing the reboiler heat and decreasing the feed preheat to compensate, so that the reflux rate remains constant.

The detailed procedure for these tests appears in Section 500.0.

306.4.2 *Maximum Operational Capacity* is the maximum capacity at which acceptable separation is achieved in stable operation. A drop in separation efficiency to a level where it is no longer acceptable is usually associated with the onset of flooding. However, in vacuum distillation, excessive entrainment will usually cause an efficiency drop well before the column becomes inoperable due to flooding in trays or packing. To find the maximum operational capacity, a series of tests is recommended at increasing feed rates, with the boil-up to feed ratio and reflux ratio held constant. A performance curve can then be plotted to find the maximum operational capacity.

306.4.3 *Minimum Operational Capacity* is the minimum capacity at which acceptable separation is achieved in stable operation. A drop in separation efficiency at minimum throughput for packing could be the result of liquid/vapor maldistribution from the distributor or within the packing, or poor wetting of the packing. For trays, a drop in separation efficiency at minimum throughput can be due to liquid/vapor maldistribution, excessive weeping of liquid through the tray bubbling area, leakage through or around trays, or pulsation of vapor flow. In any case, the procedure would be to decrease the reflux, boil-up, and feed by the same proportion until the separating action of the column falls off acceptable limits. Then, perform the test just above the minimum determined throughput value. In cases where the minimum throughput is at a value too low to meet the design separation, the recommendation is to conduct a series of tests at diminishing feed rates, with reflux and boil-up in proportion. A performance curve can then be plotted from which the minimum throughput for acceptable separation may be determined. Of course, while maintaining minimally acceptable liquid and vapor rates in practice, feed and product rates may be decreased further, even to zero.

307.0 Energy Consumption

A frequent objective of column testing is to determine operating conditions that minimize energy consumption. Two general types of tests are recommended:

Comparison of energy consumption at a fixed separation for various operating conditions. For example, in cases where feed enthalpy is significant, reflux ratio and reboiler duty can be compared for different feed temperatures, while maintaining fixed feed composition and product purities. Be aware, however, that special design considerations are necessary to ensure flashing feeds are distributed properly in both tray and packed columns.

Comparison of energy consumption with reduced product purities. At a fixed feed rate and composition, the reflux ratio and reboiler duty should be gradually reduced. A plot of product purities versus energy consumption can then be developed. The economic advantage of reduced energy consumption can be compared with the economic penalty of reduced product purity.

After obtaining a few points of data, a computer simulation may be verified or calibrated. The simulation can then be used for detailed optimization studies.

308.0 Pressure Drop Restrictions

Measurements of differential pressure (or pressure drop) across the entire column, or preferably across sections of the column, are important for most types of column testing. Frequently, the most revealing measurable distillation process variable, pressure drop, may also be a critical process variable, such as in columns handling thermally unstable materials, because of its effect on column bottom temperature. In addition, pressure drops are required when capacity tests are made (Refer to Section 502.1.2), and are invaluable when one is attempting to locate and evaluate sections of a column suspected of not operating properly [7].

For a number of reasons, installed pressure drop measurement devices can sometimes deliver false information. Therefore, pressure drop data should be measured using local pressure gauges.

309.0 Data Collection Requirements

309.1 Process Operating Data

The data to be collected for any of the tests described above include all information needed to make overall material, component, and enthalpy balances around the column. Test planning should include a complete list of all such data to be collected. The use of data historians simplifies recording the data. A record of critical data points at intervals ranging from 1 to 10 min is a valuable aid in determining when steady state is reached. Some data are best recorded by strip-chart recorders [7], data historians, or a distributed control system (DCS). If data must be recorded by hand, a prepared data form minimizes collection errors and time [5].

The number of streams to be sampled, the frequency of sampling, and methods of analysis should be evaluated as a part of test planning. Taking more data is always better than taking less. Data such as ambient conditions, battery limit utility conditions, and control valve positions may prove invaluable later in problem analysis. Process and mechanical specifications for equipment should also be made part of the test record.

309.2 Gamma Scan Data

A gamma scan can provide better detail on the actual hydraulic condition of the column. Gamma scans should be done at two different operating conditions or production rates to determine if any limits not anticipated per the hydraulic analysis are being prematurely reached or approached. If the column is exhibiting problems at high rates and not at low rates, if possible, perform a scan at both the stable low rate conditions and at the unstable high rate conditions. Scan the entire column and not just where someone thinks the problem may exist. Most of the cost in gamma scanning is in the setup. The method for scans differs from trayed columns to packed columns.

309.2.1 Trayed columns typically should have an active area scan and at least one downcomer scan. The active area should be scanned directly across, either perpendicular or slightly angled to the flow path. The downcomer scan should be parallel with the outlet weir (see Figure 4).

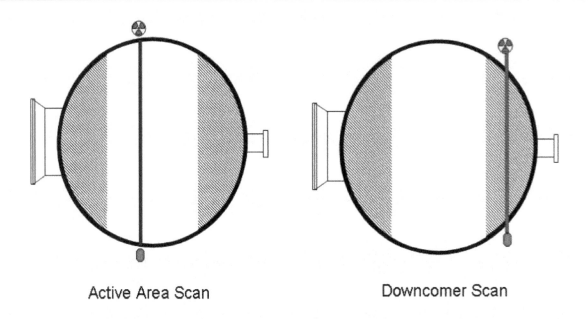

Active Area Scan   Downcomer Scan

Figure 4. Scan of Trayed Columns (view from above)

309.2.2 Packed columns must be grid scanned. This requires at least four passes (see Figure 5 on next page)

309.2.3 Scan Preparations

The following is a minimum list of preparatory items required before scanning.

1. Send the vessel layout drawings to the gamma scanning contact. This information will help determine the intensity of the source required, due to vessel wall thickness and type of internals. It is also helpful, when preparing for a gamma scan, to know the location of platforms and ladders, and have a nozzle orientation sketch.

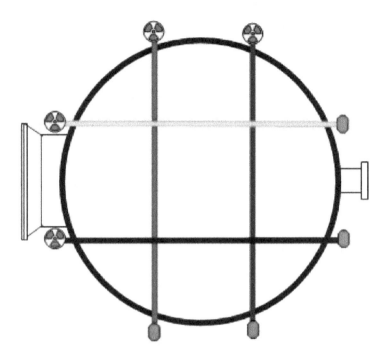

Figure 5. Scan of Packed Columns (view from above)

2. Make certain that the column is accessible to the gamma scanning equipment. Discuss the accessibility requirements with the gamma scanner.

3. Make certain that all required radiation permits are obtained. Different countries and different plant sites will vary in this requirement, even within the same company. This detail can add considerable time to the preparation.

4. Planning and precautions must be taken when gamma scanning will occur near nuclear-based instrumentation, such as nuclear level gauges, and UV-based flame scanners or detectors in flame control systems. While limited measures can be taken to avoid or reduce this possibility, it is recommended that these instruments be placed in manual-mode or standby during the scan.

5. Ensuring the best unbiased analysis dictates that opinions about the column problems not be shared with the scanner until after the analysis is presented. However, this should not prevent discussion with the scanner about what information is being

sought in case choices exist for how best to extract the relevant data. The goal here is to obtain another objective opinion based on the gamma scanning data only.

310.0 Conditions of External Streams

Conditions of external streams must be known to allow computation of overall and component material and heat balances.

310.1 Overall and Component Material Balances

The flow rate, temperature, pressure, and composition of the feed stream, distillate product stream, bottoms product and side-stream products must be obtained. For two-phase feeds, the temperature and pressure conditions at the last point where the feed is single-phase should also be obtained. Additionally, flow rate, temperature and composition are required for the reflux liquid stream returning to the top of the column for any circulating reflux in and out of the column, and for any special streams, such as the solvent in extractive distillation, or the entrainer in azeotropic distillation. Lastly, the pressure and temperatures of the overhead vapor should also be determined. However, where one such variable is theoretically redundant, it may be used as a check.

310.2 Overall Enthalpy Balance

To confirm the overall enthalpy balance, the duty added to or removed from the system must be determined by measuring the flow rate, temperatures and pressure (where vapor exists) of the heating or cooling fluids from exchangers, such as reboilers, condensers, feed preheater, etc., passing through any of the enthalpy envelopes shown in Figure 6.

These duties, together with the enthalpies calculated for the feed and products, will permit checking the overall enthalpy balance. Radiation and convection losses are usually negligible, except in special cases, such as in poorly insulated columns. Vacuum columns are another example of this exception since the mass flow rates are small compared with the external surface area of the equipment.

311.0 Internal Temperatures

Taking temperature measurements at several points within the column can be extremely useful in calculating internal flows, and in analyzing the performance of the column.

311.1 Heat Balances

If heat balances are to be hand calculated, the temperatures at points where the maximum or minimum flows occur are required. In trayed columns, the temperatures of the liquid and vapor on the trays also provide excellent insight (Refer to Section 603.2). This allows for a heat balance where the envelope intersects the column at these points, so that the internal flows may be calculated.

311.2 Internal Profiles

As discussed in Section 603.2, the calculation of internal flow rate and temperature profiles is much easier, and usually more accurate, when a distillation computer simulation program is used. Having temperature measurements at several points within the column to be compared with the calculated profiles is still beneficial. However, measured temperatures where maximum or minimum flows occur would not be necessary.

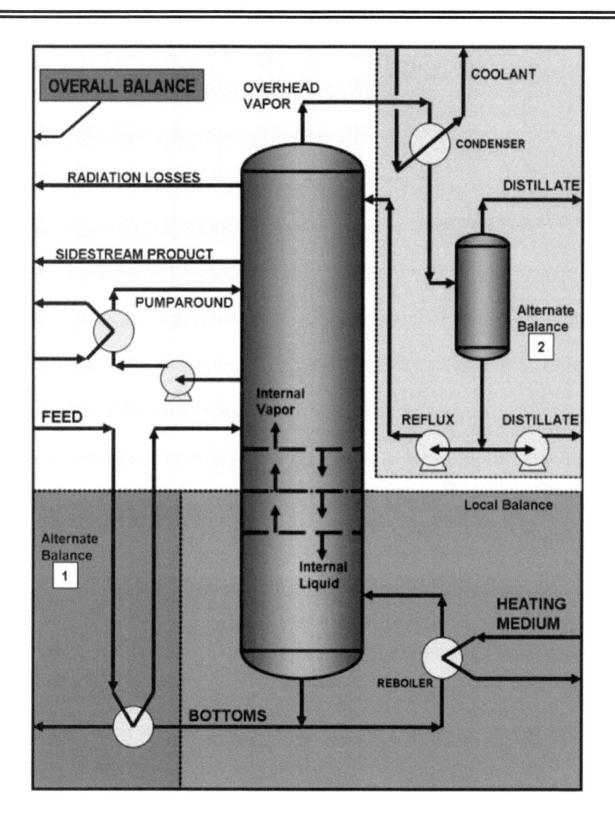

Figure 6. Enthalpy Balance Diagram

In tests where efficiency is to be determined, and where a critical analysis of the separation performance is desired, a carefully measured temperature profile of the whole column is advantageous. Temperature varies with composition and pressure, and so an accurate knowledge of the temperature profile can sometimes provide insight into the composition profile. Ideally, the temperature points should be strategically placed to cover the zones of maximum temperature change. In multicomponent columns, where moderately large amounts of non-key components are present in the feed, the composition of the non-key components change rapidly at the feed point, and near the bottom and top of the column. Consequently, these regions may have large temperature changes.

312.0 Internal Samples

It is frequently difficult to obtain samples of vapor and liquid from within the column and it should be attempted only if absolutely necessary to achieve the test objectives. Obtaining a representative sample of the vapor from a large column is unlikely. Relying on liquid samples is strongly recommended and one should obtain the average vapor composition leaving a stage by calculation. Samples taken above and below the feed point may be desirable for simplifying interpretation of test results.

312.1 Internal Samples for Efficiency Checks

Where practical, internal samples (preferably liquid) should be withdrawn from several points in the column sufficient to establish a complete concentration gradient along the section in question. This will allow for accurate efficiency determination (see Section 405.0).

312.2 Internal Samples for Overall Performance

In tests for maximum and minimum throughputs, and for overall column efficiency, internal samples are not required, since internal compositions can be estimated accurately enough to calculate fluid densities and enthalpies. However, this point should be checked by a preliminary heat balance.

313.0 Pressure Profiles

Pressure-drop measurements are always desirable, especially in packed columns. Accurate differential pressure meters or manometers are required (see Section 404.0).

314.0 Data Requirements-Physical Properties

314.1 Test Mixtures

The test mixture must be suited to the column being tested. In particular, the relative volatility and composition must be such that extremes of purity are not reached at either end of the column. Such extremes can exaggerate the effect of errors in composition due to the resulting pinches. Moreover, where one component is in low proportion, it can be largely held up in the column, reflux drum, or sump, leading to difficulty in reaching steady state. Of course, the use of a mixture for which the column was intended would be most convenient, but because of the reasons cited above, this may not be practical. Descriptions of test mixtures suited to various heights of columns and operating pressure are available [9,10].

314.2 Essential Data

Essential physical property data must be available for calculating and interpreting the performance data. Essential data of vapor and liquid streams over the range of column conditions include densities, molecular weights, and latent heats and heat capacities, or enthalpy correlations. Heats of solution should be included when this property affects stream enthalpies. Vapor-liquid equilibrium (VLE) data are required for graphical stage/tray counting or computerized column simulation. Accurate VLE data or correlations are also essential for meaningful determination of theoretical stages.

Physical property data are available in several databases sponsored by AIChE and other major technical societies, worldwide. Many texts also are also devoted to the estimation of these properties.

315.0 Auxiliary Data

Other physical properties useful in developing and testing correlations, and for analyzing the performance data, include viscosity, diffusivity, and surface tension.

316.0 Test Procedure Documentation

The final phase of test planning should be a written document containing all pertinent test information. A complete document would include a summary of test objectives, manpower requirements, essential physical property data, pretest calculations, data collection procedures, sampling schedules, test sequence and any special safety and environmental considerations. Reference should be made to the relevant standard sampling procedures, equipment data, and safety and environmental regulations. A summary report for management could include only the test objectives, manpower requirements, and safety and environmental considerations. Finally, a session should be held with the unit operators to fully explain the test and their duties and responsibilities during its operation [5].

400.0 METHODS OF MEASUREMENT AND SAMPLING

401.0 System Controls and Operating Stability

Reliable measurements and samples can only be obtained from a column operating smoothly within its design operating range. Any deviation, such as pressure swings in a column, will result in vapor/liquid rate fluctuations, and temperature and composition changes. Consequently, before rigorous testing begins, any serious deficiencies that could interfere with the steady operation of the column should be diagnosed and corrected.

A few of the problems frequently encountered are as follows:

- Changing feed compositions
- Improperly tuned control loops
- Changing steam or cooling water supply pressure
- Oversized or sticking control valves
- Surging reboiler
- Liquid level controller failure
- Pressure swings due to ambient temperature changes
- Poor calibration of on-line analyzers

402.0 Measurement of Temperatures

Any reliable thermometer may be used to measure temperature, but thermocouples or resistance temperature detectors (RTDs) placed in thermowells are preferred in commercial installations due to their ruggedness and availability. Before use, all temperature-measuring devices should be checked and calibrated in the temperature ranges to be used. RTDs are usually more accurate than thermocouples.

402.1 Accuracy

Temperatures should be measured accurately enough that the maximum cumulative error of the heat quantities calculated from temperatures and flow rates is less than 5%. For example, in measuring the heat removal from a condenser by water, temperature measurements within a fraction of a degree are sometimes required. A relatively accurate temperature difference can be obtained by connecting two thermocouples in series opposing each other. Differential temperature accuracy can also be improved by using temperature sensors manufactured in the same batch, as they will typically have the same calibration curve.

402.2 Errors

Errors in temperature measurement can still occur even if the sensing elements are accurate.

> 402.2.1 Recommended techniques and practices should be followed to eliminate errors due to conduction and radiation [11]. Thermowells must have an adequate length of insertion into the process fluid being measured. Both the thermowell connection and the external surface of the column

near the thermowell must be well insulated. Good thermal contact between the sensor and the thermowell is also important.

402.2.2 Temperatures inside the column are sometimes measured by inserting the thermowell into the packing bed, and on the tray decks, or in the downcomers.

402.2.3 For a packed column, installing two or more thermowells at each sensing level is advisable, particularly for larger diameter columns. More measurement points are especially desirable when liquid distribution quality is a concern. These points are equally spaced around the perimeter of the column, but each is inserted into the packing to a different depth. The average reading of these sensors is used as the measured temperature at the subject sensing level. The differences between the average and individual readings can be used as a qualitative measure of liquid and vapor distribution across the column. Liquid temperatures between packed beds are obtained in the liquid collector rings or trays, in the downcomers from the liquid collector trays, or in the liquid re-distributors.

402.2.4 For a trayed column, liquid temperatures should be measured in the downcomer, or in the seal pan. For trays with truncated downcomers, the thermocouples should be placed in the bottom of the downcomer where clear liquid exists.

402.2.5 Accurate vapor temperatures in columns are difficult to obtain, and should be avoided unless no other options exist. The accurate measurement of vapor temperatures calls for probes that permit vapor to flow past them, but are effectively shielded from liquid impingement and other sources of conducted or radiated heat to or from the sensing point. Multiple probes at each vertical location, or a single probe with externally variable lengths of insertion into the column, are recommended.

402.2.6 Special attention should be given to measuring the top column temperature. The reflux temperature is usually sub-cooled and will affect measurements of liquid temperature in the reflux false downcomer. The liquid in the downcomer from the top tray is a good location for the top temperature. However, a vapor temperature measurement in the column dome or vapor line leading to the condenser is more accurate.

402.2.7 All thermowells should be inspected before a test is made in an installation where deposits may occur that could affect response time.

402.2.8 In a thermosyphon or other types of reboilers, where the outlet line to the column contains two fluid phases, quite different temperature readings will be noted, depending upon the vertical location of the thermocouple in the outlet line. This is due to the continual flashing of the material as the hydrostatic head is reduced in the rising fluid. To get a meaningful fractionator temperature, it is recommended to place the thermocouple in the liquid in the column bottom sump at the same elevation as the lower liquid level nozzle. A thermocouple directly at the reboiler outlet will be useful only for observing reboiler performance.

402.2.9 Special attention is required to determine the correct enthalpies of feeds to columns. If a liquid feed from a preceding unit flashes at column pressure, its temperature must be measured at a location in the feed line or preceding unit where single-phase liquid flow exists. If the feed is converted from a liquid to two phases in a feed pre-heater, the temperature of the liquid entering the pre-heater, plus the heat input to the pre-heater per unit of feed, are both required to calculate the feed enthalpy.

402.2.10 Some control systems use "pressure compensated" temperatures, rather than raw values, in an attempt to hold a constant composition profile, rather than a constant temperature profile. To evaluate column performance, it is important to know the actual temperature and pressure values, and not just the compensated value.

403.0 Measurement of Flow Rates

The rate of flow may be measured by means of instruments, such as orifices, rotameters, venturi, displacement, vortex, or mass flow meters, or by direct volume and weight measurement. Whenever possible, the instruments should be calibrated in place with the fluid to be measured, and at the temperature and pressure (gas measurement) at which it is to be used. Meters must be accurate enough for the material and heat balance requirements (see Sections 602.0 and 603.1).

403.1 Orifice Meters

Orifice meters have been extensively investigated, and their performance can be accurately predicted for a variety of conditions. Details concerning the construction, installation, and calibration of orifices and nozzles can be found in *Fluid Meters-Their Theory and Applications* [12]. An alternative procedure is outlined in *Perry's Handbook* [1].

The reading of orifice or venturi-type meters is affected by the density of the flowing fluid. If the density differs from that for which the meter was calibrated, the reading, if in volumetric units, must be corrected as follows:

$$\text{Volumetric flow rate} = (\text{Volumetric reading}) \times \left(\frac{Design\ Density}{Actual\ Density}\right)^{0.5} \quad \textbf{(403.1.1)}$$

If the meter is also designed to read volumetric flow corrected to a standard temperature and pressure, a further correction is needed if the density differs from the design density of the meter. The combined corrections give:

$$\text{Volumetric flow rate} = (\text{Volumetric reading}) \times \left(\frac{Design\ Density}{Actual\ Density}\right)^{0.5}$$

$$\times \left(\frac{Design\ Density\ at\ Standard\ Cond.}{Actual\ Density\ at\ Standard\ Cond.}\right) \quad \textbf{(403.1.2)}$$

Viscous materials may require an additional correction for viscosity [12].

Failure to check the orifice plate to verify that it is the proper size and in good condition can ruin the test results. Recalibrating all flow meters for the purposes of the test run is recommended wherever possible to measure the mass flow rate directly. Direct reading instruments that provide immediate heat balances during testing will help to identify errors in flow rates.

When vapor flow rate is measured in high-pressure services, the correction due to the compressibility also needs to be considered [1].

403.2 Rotameters

When rotameters cannot be calibrated in place, data can be accessed from generalized charts prepared by manufacturers. The theory developed by Head [13] may be used to convert the calibration from one fluid to that for another, and to correct for variations in temperature and pressure.

403.3 Vortex Flow Meters

Vortex flow meters, also known as vortex shedding flow meters or oscillatory flow meters, measure the oscillations of the downstream vortexes caused by a barrier, which is usually cylindrical in shape, placed in a moving stream. The frequency at which these vortices alternate sides is essentially proportional to the flow rate of the fluid. Inside, atop, or downstream of the shedder bar is a sensor for measuring the frequency of the vortex shedding. This sensor is often a piezoelectric crystal that produces a small, but measurable, voltage pulse every time a vortex is generated. Since the frequency of such a voltage pulse is also proportional to the fluid velocity, a volumetric flow rate can be obtained using the cross-sectional area of the flow meter. Vortex flow meters can be used for both gas and liquid flow measurements, but are not recommended for applications with corrosive or dirty liquid.

403.4 Coriolis Flow Meters

A Coriolis flow meter can directly measure mass flow and fluid density. Coriolis measurement can be very accurate, no matter what type of gas or liquid is measured. Since it directly measures the mass flow rate, the same measurement tube can be used for other fluids without recalibration.

403.5 Magnetic Flow Meters

More commonly called a mag meter, a magnetic flow meter does not have any moving parts and is ideal for wastewater applications or any dirty liquid that is conductive or water-based. Technically, it is an electromagnetic flow meter and works by applying a magnetic field to the metering tube/pipe, which results in a potential difference proportional to the flow velocity perpendicular to the magnetic flux lines.

The magnetic flow meter requires a conducting fluid, such as water, that contains ions, and an electrical insulating pipe surface, like a rubber-lined steel tube. Though they will generally not work with hydrocarbons, distilled water and many non-aqueous solutions, they work well where low pressure drop and low maintenance are required.

403.6 Pitot Tube (or Annubar)

A Pitot tube is a device that measures fluid static pressure. The fluid flow rate within a pipe is calculated from the difference between the stagnant pressure and the static pressure of the fluid. An Annubar has several small openings placed in a hollow "bar" across the width of a pipe. The front, or upstream, openings measure the average stagnant pressure, and the openings in the back, or downstream, side measure the average static pressure. The fluid must be a single phase and the Pitot tube or Annubar should never be used in fouling service.

403.7 Direct Volume or Weight Measurement

Whenever practical, the fluid in tanks or gas holders should be weighed or measured volumetrically at specified intervals of time as a check on other flow measurements, and a plot of several measurements recorded versus time. In situations where no steam meters are present, or when checking their accuracy is desirable, steam condensate can be collected into a large container. When correctly timed, this technique gives an accurate measure of the steam flow rate. To minimize condensate flashing and safety risks, a condensate cooler should be used.

404.0 Measurement of Column Pressure Drop

Differential pressure drops across sections of packed bed, and trays are used to monitor, and sometimes control, the vapor rate through the columns, as well as to monitor column conditions. Pressure taps should be provided so that the performance of each bed or section of trays can be checked as needed.

404.1 Instruments

Manometers are the most convenient instruments to install for a temporary measurement of the pressure drop. They are portable, easy to install, and have a wide range of operation. They should be equipped on both sides with seal pots.

Differential pressure cells are recommended for high-pressure columns and for permanent installations where a remote reading is desired. The cell can be connected to the column with or without the use of seal pots. They should be calibrated before use in a test. The lines connected to the differential pressure cells should be purged with inert gas to prevent condensate from collecting in them.

Compared to pressure drops across trays, pressure drops across packed beds are usually quite small, generally ranging from 0.1 to 1 inch of water per foot of packed height (8.3 to 83 mm of water per meter). Therefore, to evaluate packing performance, care must be taken to obtain readings that are accurate and reproducible. Differential instruments must be used.

Pressure differences measured between separate gauges are not acceptable when the differential pressure is not a significant fraction of the absolute column pressure. If a pressure gauge must be used, either the same gauge must be moved and used for all readings, or the gauges must be calibrated so that differential pressure can be accurately measured.

404.2 Pressure Taps

Column pressure taps must be located where no danger of submergence below a liquid level exists. Where applicable, it is preferable to have gas or vapor-filled lines leading to the pressure differential sensing device. Instruments that use liquid-filled capillaries to transmit pressures from diaphragms to the differential pressure sensor are subject to pressure measurement errors. The liquid head of the fill fluid is compensated for when calibrating the measurement. The errors are especially acute in systems with large elevation differences due to the density of the fill fluid varying with the ambient temperature. The instrument tubing and taps must be clear of obstructions, such as polymer material or liquids that will interfere with the collection of accurate pressure data.

Where practical, the differential pressure sensing device should be installed high enough so that gas- or vapor-filled lines leading from the column to the sensing device drain freely back into the column. It is also very important to employ lines of the proper size to avoid vapor condensations and capillary liquid hold-ups. One-half to one-inch (13 to 25 mm) diameter lines are usually recommended. Low spots in these lines that can collect condensate should be avoided. Line diameters should be adequate so that pressure drops due to gas purging are negligible, and so that the lines are free draining. Provision must be made to drain or blow condensate from the sensing system during startup and normal operations without interfering with the performance of the column. When vapors condensing and freezing in the sensing lines is a possibility, the lines must be heat traced or jacketed, or continuously purged with an inert gas or vapor. Pressure tap lines should also be purged or filled with a suitable liquid or gas if the material in the column is corrosive or toxic.

As a practical guide, the taps used for the pressure drop measurements are recommended to have:

1) Large and constant L/D ratio (depth of tap/tap diameter) of at least L/D>2 to make sure that the flow within the cavity (tap) is fully developed.

2) A small ratio of tap diameter to pipe/column diameter to minimize the effect of tapping on external flow stream.

3) For taps with a small L/D ratio (<2), a wide cavity behind the tapping is suggested to reduce the error [14].

The second type error will occur if the static pressure tap is not flush with the wall, or protrudes into the column. This might happen if the pressure taps were mounted incorrectly, or if the surface of the column wall was eroding or ablating. Protruding taps will disturb the boundary layer near the wall and result in static pressure measurement errors. The pressure obtained by the protruding tap is smaller than the "true" wall static pressure [14]. The larger the protruding length, the bigger the potential error. Therefore, protruding taps should be avoided for column pressure drop measurements.

404.2.1 *Gas Purge:* This technique is the preferred method for keeping lines free from column vapors. The purge flow rate must be steady and low to avoid line pressure drop. If a high purge flow rate is required, it is important to maintain the same flow rates for two lines connecting to the same pressure cell. A rotameter or bubbler is used to indicate the flow rate, which is adjusted by the needle valves. A pressure controller is installed to hold the purge gas pressure constant at a value higher than the column pressure. As an extra precaution, seal pots (Section 404.3) will allow condensed vapors from the column to drain back and dampen pressure fluctuations. Refer to Figure 7 for a typical installation.

In high-pressure columns, the reading will include the static head of high-density gas in the pressure taps and the static head of high-density vapor in the column. The static head between the pressure taps can be compensated for by taking the zero reading at the operating pressure, or by calculation [15]. It is important to blow out the pressure sensing lines during start-up and operations without disturbing the performance of the column. Gas purge systems are not recommended for systems that cannot tolerate build-up of a non-condensable in the condenser.

404.2.2 *Liquid Purge:* The use of liquid may sometimes be more convenient than a purge gas, but the basic principles of the purge still apply. In addition, the installation should be made to permit the complete elimination of air bubbles above the liquid in the manometer to the seal pots on the column. The selected purge liquid should have a constant known specific gravity. The zero reading is also made with purge liquid flowing at the desired rates with the side vents on the seal pots open. See Figure 8 for a typical installation.

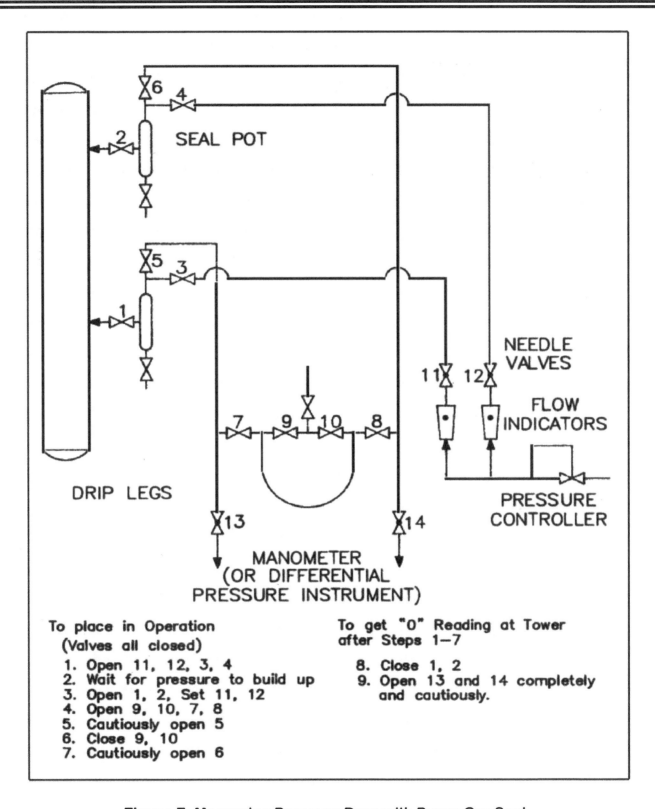

Figure 7. Measuring Pressure Drop with Purge Gas Seal

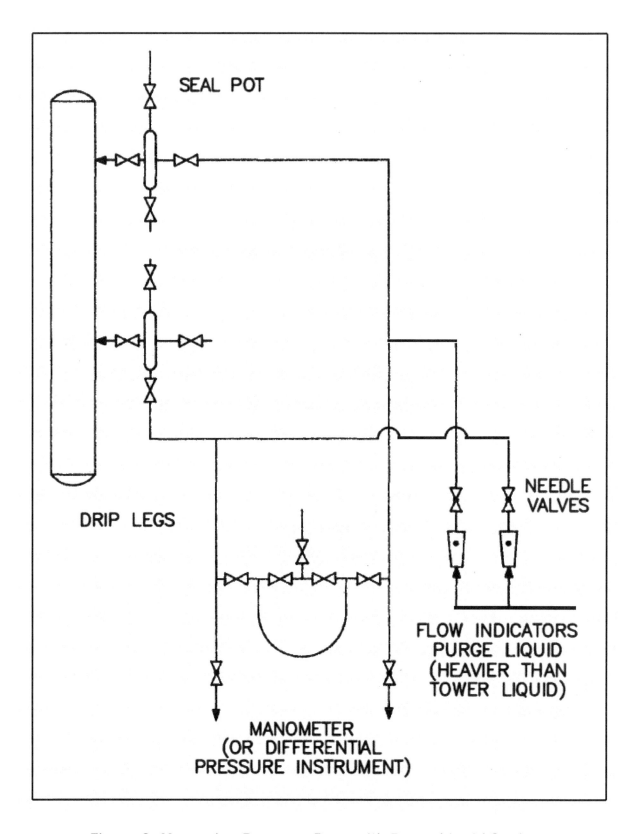

Figure 8. Measuring Pressure Drop with Purge Liquid Seal

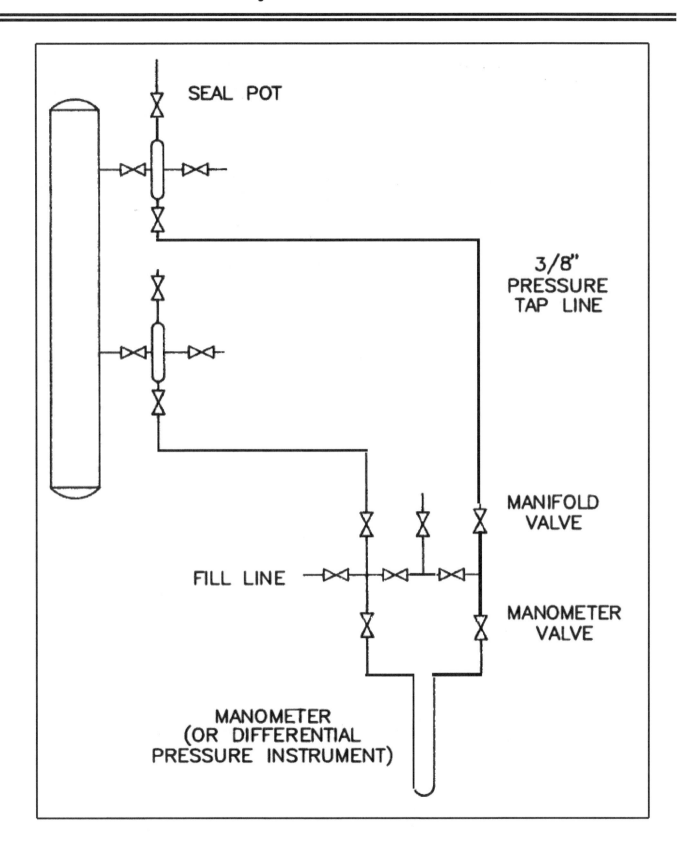

Figure 9. Measuring Pressure Drop with Static Liquid Seal

404.2.3 *Static Liquid Seal:* A static liquid seal is used on columns where the process fluids must be kept out of the instrument lines, but where a continuous purge is impractical. The sealing liquid then should be immiscible with, and heavier than, the column liquid or condensate. The installation should permit complete elimination of air bubbles in the instrument and lines to the seal pots, which are filled to the level of the column connections. The zero reading is made with the vent valves open. Refer to Figure 9 for a typical installation.

404.2.4 *Vapor-filled Lines:* If the column vapors are not condensable at ambient temperature or if the lines can be steam- or electrically-heated to prevent condensation, no sealing fluid is necessary. The seal pots, as shown in Figure 9, for the gas purge are still recommended, however, to trap column liquid that may surge into the line.

404.2.5 *Pressure-tap Line:* A flexible high-pressure hose with rapid disconnect pressure fittings is useful for many pressure drop measurements.

Dynamic pressure drop shown in Table 1 is the typical value for structured packings across the range of useful capacity. As seen from the table, as operation pressure increases, the static head becomes significant compared with the dynamic pressure drops. Making static head corrections is necessary when pressure drops are measured and reported.

The column pressure is generally measured at the top of the column. The existence of static vapor head in the column and pressure drops will make the pressure at the bottom of the column higher, especially for high-pressure trayed columns. To calculate the column local pressure, the pressure drop and the static vapor head need to be added to the pressure measured at the top.

404.2.6 *Vapor Static Head and Column Pressure:* The differential pressure between two locations in a distillation column has two components—static head due to the weight of the vapor between them, and dynamic pressure drop due to the resistance of column internals to the flow. Most published pressure drop data does not correct for the vapor static head, or mention if the correction had been made.

The static head is usually less significant in a trayed column as the minimum dynamic pressure drop for a tray is often about 1 inch H_2O/tray or higher (25 mm H_2O/tray). However, the static head may cause serious errors/problems for a packed column, especially for low-pressure-drop packings. Table 1 compares dynamic pressure drop and static head for some of hydrocarbon systems measured by FRI (Fractionation Research, Inc.).

System and Pressure	Dynamic Pressure Drop	Static Head
	inch H_2O/ft, (mm H_2O/m)	inch H_2O/ft, (mm H_2O/m)
o/p-xylene, 100 mmHg	0.01-0.84 (0.8-70)	0.01 (0.5)
Cyclohexane/n-heptane, 5psia (0.34 bar)	0.02-0.54 (1.7-45)	0.01 (1.1)
Cyclohexane/n-heptane, 24 psia (1.65 bar)	0.05-1.02 (4.0-85)	0.06 (5.0)
Iso-butane/n-butane, 165 psia (11.4 bar)	0.01-0.60 (1.0-50)	0.35 (29)

Table 1. Comparison of Dynamic Pressure Drop and Static Head

404.3 Seal Pots

Seal pots are recommended in all connections from column to pressure-tap lines. A suggested size is an 8 inch (about 200 mm) length of 2 inch (50 mm) pipe. The line to the column should be 3/4 inch (19 mm) or larger. The design should allow for the connecting line to absorb the total temperature gradient from process fluid to ambient temperature. The other connections should be 3/8 inch (9 mm) or larger. The materials of construction should be the same as the materials in the column, or equivalent materials must be selected for valves, nipples and pressure-tap lines that can withstand the column operating conditions.

404.4 Leakage Check

Any leakage in the tubing and fittings between the differential pressure cells and pressure taps will greatly affect the pressure drop results. It is recommended that all tubing and fittings be pressure-tested for leakage before taking any actual measurements.

404.5 Accuracy

In well-designed, properly operated installations, the accuracy of measurement depends largely on the readability of the instrument. Differential heads may be measured within 0.1-inch (2-3 mm) of manometer fluid, particularly if an inclined manometer is used. If a static liquid seal (Section 404.2.3) is used, the formula in Section 5-6 of *Perry's Handbook* [16] should be used to account for the change of sealant level in the lower seal pot. Each liquid may dissolve in the other to some extent. The densities should be determined after the two fluids have been mixed, and at the temperature at which the readings will be taken. Tower liquid or condensate may replace seal liquid in the seal pot where the level dropped. Here, the formula should be corrected by making a pressure balance on each side of the manometer system.

Most differential pressure cells have an accuracy of 0.25% of full scale or higher. High stability and high accuracy pressure cells can offer errors as low as 0.05% of full scale, depending on the model. Although more expensive than general purpose differential pressure cells, they may be the only option if high precision is required. No matter the accuracy, all differential pressure cells need to be calibrated before taking any pressure drop data.

405.0 Sampling Procedure

The success of a performance test greatly depends on proper sampling techniques. Usually, each sample must represent the stream from which it is taken and it must be delivered to the analyst without loss or contamination. Because proper sampling requires great care, it should be carried out under the direct supervision of the engineer responsible for the performance test.

Note: Sampling should be done only by trained personnel using proper personal protective equipment, as samples may be under pressure, toxic, hot to the touch, or may flash or react with air or water.

405.1 General

The device used to analyze concentration should be calibrated with known test mixtures of the fluids used in the experiment. The very low and very high concentration ranges should be checked for any decrease in precision and accuracy of the measuring device. Duplicate samples should be run by analysts involved in the test work to obtain an error analysis record.

Duplicate samples should always be provided to verify the sampling and analytical technique. Samples should be withdrawn at such a low rate that the area being sampled is not starved or disturbed to upset the steady state compositions. Precise temperature and pressure measurements should be made at the sample points for temperature-pressure- composition checks and correlations.

405.2 Selection of Sampling Points

405.2.1 Samples of streams external to the column (feed, reflux, products) should be taken where the stream is all liquid or all vapor and well-mixed. Liquid samples (rather than vapor) give better accuracy. If vapor samples must be used, they are best routed directly to the analyzer via tubing. Otherwise, they need to be collected in sample vessels. Tubing connections must be heated to prevent condensation. Avoid locations where stratification or phase separation in the line may exist, such as at the outlet of a condenser. Instead, select points where the fluid is well-mixed, such as following pumps or control valves, but not where vaporization due to flashing exists.

405.2.2 For packed columns, samples of liquid from packed beds can be obtained from collectors directly under the packing support plate, or from a liquid collector tray below the bed in question. A cross-sampler can be placed below the support plate to get a representative liquid sample across the entire column section [17]. Good practice calls for several liquid collector sampling points around the column perimeter or across the column to help diagnose packing performance problems. If the amount of entrapped vapor plus the vapor density are great enough to influence liquid analyses, an internal or external vapor-liquid disengagement device may be required.

405.2.3 For trayed columns, samples of liquid from within the column should be taken where the liquid is free of vapor. A preferred location is at the base of the downcomer where the liquid is mixed and essentially degassed. Even at this point, however, evidence exists that the liquid may contain entrained vapor, and it is recommended that a settling chamber be provided for the liquid at this location to allow a vapor disengaging volume and a more representative sample. A liquid sample taken at the base of a downcomer represents the composition of the liquid entering a tray. Liquid from trays without downcomers should be taken from the center by means of a settling chamber. For trays with truncated downcomers, a small sampling cup can be placed at the bottom to collect liquid samples.

405.2.4 Samples of vapor from within the column must be free of any entrained liquid. Kastenek and Standard [18] describe a vapor sampler design that has been used with some success, while Lockett and Ahmed [19] offer a modified version.

405.3 Sample Connections

To reduce the need for purging large amounts of material, the sample lines should be smaller than 1/2-inch (13 mm), preferably 1/8 to 1/4 inch (3-6mm). They should be as short as possible and, for vapor samples, should be free from low sections where liquid may be trapped. If the permanent block valve at the sample outlet is too large to give the necessary control during withdrawal of the sample, a small valve should be installed downstream from the block valve. This will be subsequently called the sample valve.

Valves and lines for sampling vapor at its dew point must be heated to prevent condensation. Sampling materials that freeze at elevated temperatures will require heat traced or jacketed sampling lines and valves. In cases where a cooling coil is required, it should be located between the block valve and the sampling valve. Generally, this coil will be 1/4 inch (6mm) tubing in an open container filled with coolant. With liquid samples that do not freeze under ambient conditions, the coil should permit cooling to 80°F (27°C) with 75°F (24 °C) water. Additional equipment (pressure gauges, reservoir for liquid used for displacement) is required for pressure sampling.

405.4 Containers

405.4.1 Two types of containers are used depending on the boiling point and toxicity of the material. Nontoxic materials in which all components boil above 200°F (93°C) are usually sampled in open containers as liquids. When necessary, the samples should be withdrawn through a cooler, with Pyrex bottles or metal cans of corrosion-resistant material preferred. Toxic materials, and those that boil below 200°F (93°C), are sampled either as a gas or liquid into cylinders or sample vessels. In addition, the effect of the atmosphere on the sample must be considered; for example, hygroscopic liquids must be sampled in closed containers, and freezing liquids may require that the sample container be heated during sampling.

405.4.2 Most companies have standards for construction and testing of containers. However, without these standards, the regulations of the Surface Transportation Board of the United States, housed within the U.S. Department of Transportation, are recommended. These regulations also specify sizes and methods of packing and labeling if transporting samples by common carriers is necessary. Cylinders and sample vessels must have two connections, at opposite ends of the container and be closed with valves. For high-pressure sample vessels, the valves are preferably diaphragm-packed globe valves, as used on commercial gas cylinders, with the seat side connected to the container to minimize danger of leakage. For low-pressure gas cylinders, small needle valves are suitable. Small glass containers are usually used for low-pressure gas samples if the analytical method does not require large samples.

405.4.3 The size of the container should be large enough to permit at least two accurate analyses.

405.4.4 The containers should be inspected for cleanliness and leaks before use. Sample vessels should be pressure-tested if the sample is to be drawn and stored under pressure.

405.5 Sampling of High Boiling Materials

405.5.1 Sampling of a nontoxic stream that has no component boiling lower than 200°F (93°C) can normally be done in open containers that are immediately stoppered. For a vapor sample, it should be condensed and cooled to at least 80°F (27°C) as it is withdrawn to prevent flashing.

405.5.2 The sample line should first be flushed out to a safe place (or used to rinse the container) with a volume of liquid that is three or four times the dead space of the connections. During sampling, the sample tube should extend into the bottom of the container to prevent splashing and loss of light materials. The container should not be filled entirely; about a 10% outage is recommended to allow for expansion, and it should be immediately capped or stoppered.

405.5.3 Caution should be used in handling flammable materials. These fluids often generate static electricity. Where the fluid flows through non-conducting lines or open air, high-voltage charges may build up. Therefore, it is important to ground metal sample containers to the metal at the sample source.

405.5.4 In cases where extreme care must be used in preventing loss of light material, sample vessels should be used in place of open containers, and the methods of Section 405.6 apply.

405.5.5 The above methods assume that higher than atmospheric pressures are available. Vacuum systems can be sampled by ordinary methods if the liquid column is adequately purged before sampling, and if the sample point for a liquid, or the sample condenser for a vapor is located so that a liquid column sufficient to break the vacuum is available. If this sampling method is not practical, a sample vessel, as described in Section 405.4.2, can be attached to the sample connection at the lower valve, while the upper valve is vented back to a vacuum source by way of a liquid trap. Purging through the sample vessel should be done before sampling.

405.6 Sampling of Intermediate Boiling Materials
(Approximately -50°F to 200°F or -46°C to 93°C Boiling Point).

These materials are usually sampled in the same physical state as that in which they exist at the sample source. Care must be taken to keep vapor samples all in the vapor phase. Likewise, liquid samples should be monitored so as not to allow any vaporization, since any vapor formed would be trapped in the lines leaving the container. The preferred method of sampling, either gas or liquid, is to displace them into sample vessels by liquid. An alternative method is by purging. If neither of these procedures can be used because of the nature of the materials being handled, the sample may be taken into an evacuated sample vessel.

405.6.1 *Sampling by liquid displacement:* The equipment recommended is shown in Figure 10. A suitable liquid for displacement must be chosen so that no transfer of material takes place. For hydrocarbons, brine is generally used. The sample vessel can be filled with the displacement liquid before being connected to the sampling line. This method has the advantage of purging the air from the line above the sample vessel, but this is usually small enough to be negligible. When all air has been displaced, the sampling line is purged out with the sample vessel closed off. After the vent has been closed, the procedure will vary, depending on whether a gas or a liquid is being sampled.

Gas is allowed to flow into the container at such a rate that the pressure is maintained at a value consistent with the design of the sample container, or at a rate that will not cause excessive cooling due to rapid expansion. Conditions in the sampling line and container should be such that none of the heavier constituents condense. If, during purging, drips of condensation are observed, no sample should be taken until this condition is corrected. When the desired amount of sample has been collected by measuring the displacement liquid drained from the sample container or displaced back into the reservoir, the valve at the base of the container and the sampling valve are simultaneously closed. The other container valve is then closed and the container disconnected.

Liquid is allowed to flow into the container only while pressure in the container is maintained above the vaporization point. If the temperature of the source is above 80°F (27°C), a cooling coil may be necessary to assure that no vaporization takes place during sampling. All valves are wide open except the vent valve, which is closed, and the lower container valve, which is throttled to maintain flow at the desired rate. The amount of displacement liquid removed from the container is measured and the flow is stopped when the desired displacement liquid remaining is equal to the

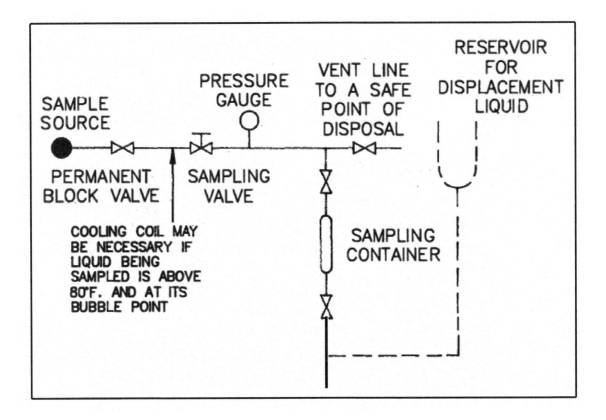

Figure 10. Sampling by Displacement

amount of outage required. The container is then closed off from the sampling line and the remaining displacement liquid forced out by the vapor pressure of the material being sampled. The amount of outage required to provide space for liquid expansion is usually 5%, but this may be modified for special cases where a significant portion of the lightest component may be transferred into the vapor space. Containers filled with liquid must never be left with both valves closed because of the danger of rupture if the temperature increases. The necessity for outage in containers filled with displacement liquid can be avoided by leaving one valve open.

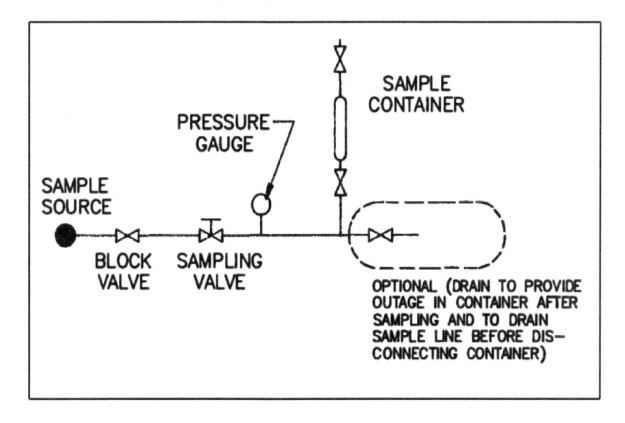

Figure 11. Sampling Liquids by Purging

405.6.2 *Sampling liquids by purging:* The equipment required is shown in Figure 11. The sample container is pressured with the sample. Then the compressed vapors and enough liquid to displace the contents of the container and system three times are allowed to escape from the top vent at such a rate that pressure is maintained in the container. Valves are then immediately closed, providing the necessary outage. Depending on the materials being handled, it may also be advisable to provide a drain for this purpose. For outage requirements and other procedures applicable to pressure containers, see Section 405.6.1.

405.6.3 *Sampling gases by purging:* The equipment required is shown in Figure 12. During purging, the sample valve is throttled to prevent exceeding the maximum sample pressure. The sample container valves are closed when at least three times the volume of the sample line and container has been purged through.

405.6.4 *Samples taken into evacuated sample vessels:* When this method is used, the sample line is purged and then the evacuated sample

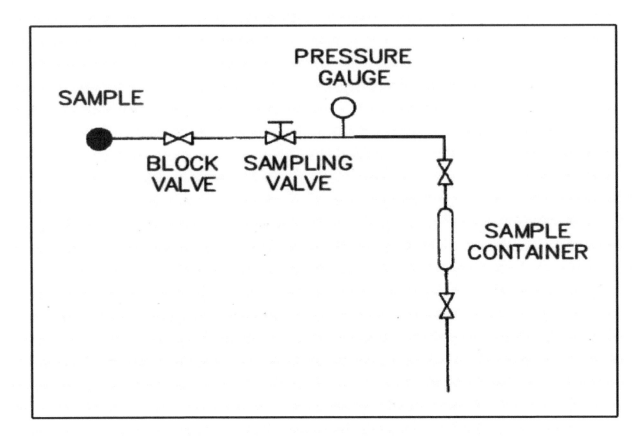

Figure 12. Sampling Vapors by Purging

vessel is attached to the sample line with suitable fittings. The inlet valve to
the sample is opened, then closed, and the sample vessel removed. Flashing
of the liquid inside the sample vessel may be tolerated as the sample may be
totally condensed, either before removal from the sample vessel or, for
lighter liquids, the entire sample may be removed from the sample vessel
as vapor. Correction should be made for the small amount of air present in
the connecting tube between the sample vessel and the sample line from
the column, although this can be eliminated if the sample vessel and sample
line can be evacuated in the field after the two have been connected.

405.7 Sampling Materials Having Boiling Points Below -50°F (-46°C)

It is always recommended that these materials be collected as gases. If a liquid stream is to be
sampled, it should be continuously vaporized by a heater placed in the sampling line. Streams
having no components that boil above -50°F (-46°C) can be easily vaporized at atmospheric
pressure without danger of leaving a small amount of the heavy components behind as liquid. A
heating coil may also be necessary for gas samples taken from sources having pressures over 300
psig (2170 kPa) to prevent cooling by expansion below the freezing or condensation point of some

constituents. The heater, in any case, should be placed between the block valve and the sampling valve. Otherwise, the equipment is similar to that required for gases (Figures 10 or 12).

405.8 Leakage Check

Any leakages in sampling systems will impair the concentration measurements. Therefore, all sample lines, fittings, and other auxiliary equipment related to sampling should be closely inspected for leakage. Leakages found should be fixed before taking any liquid/vapor samples.

405.9 Labeling and Handling the Samples

405.9.1 The sample should be tagged immediately after being taken and the following information transmitted with the sample:

1. Unit, stream sampled, sampling point

2. Sample number

3. Company and location

4. Date and time

5. Pressure, temperature, and phase of sampling point

6. A note telling the analyst how long the sample should be retained before disposing

7. An approximation of the composition of the sample so the analyst can properly prepare the analytical apparatus. If possible, the approximate bubble points of liquid samples and dew points of vapor samples should be estimated and reported to the analyst

8. Name of person conducting the test

9. Weight of sample vessel after sampling

10. Any special condition that would affect the manner in which the sample should be withdrawn; for example, if the gas was at its dew point, or a liquid at its bubble point, under the conditions of item number 5 above

11. Any hazardous materials or special handling procedures.

405.9.2 Samples should be analyzed as promptly as possible to minimize any change of composition caused by leaks. However, if storage is necessary, liquid samples should be stored in a cool place, and vapor samples at a temperature where condensation will not occur. Storage should be located where leakage will not create fire or toxic hazards. Samples may undergo change in composition due to chemical reaction, such as polymerization, oxidation, condensation or decomposition. Consequently, the samples should be analyzed quickly. If not, all possible steps should be taken to avoid chemical reaction or, at least, to estimate its rate. If the sample is inhibited, this should be noted.

405.9.3 To check for leakage, it is recommended that sample vessels be weighed immediately after sampling, and just before removal of the analyst's sample.

405.9.4 The analyst should ascertain the condition in which the material exists in the container. For gas samples, heating may be required to vaporize any condensate. For liquid samples under pressure, the liquid should be cooled below the temperature at which it was sampled and the analyst's sample should be withdrawn from the liquid phase in the container. If, during cooling of the liquid sample, the vapor pressure is below that required to remove the analyst's sample, the sample vessel should be pressurized with an inert gas such as nitrogen or the sample forced out with a displacing fluid such as brine.

500.0 TEST PROCEDURE

501.0 Preliminary

Proper inspection and testing should be made during assembly and/or before the start-up of new or revised columns. Such tests include:

- Leakage tests for liquid distributors or trays, which are usually important for bubble cap trays in very low liquid rate applications

- Inspection of rings and attachments for supporting plates and distributors to ensure they are level

Usually, it is desirable to know that the internals are assembled properly, that the unit is clean, and that the dimensions, location, and orientation of the internals are as specified in the design. See Kister [6, 20] and Taylor and Krishna [21] for details on this subject.

All control loops should be operating in a stable manner, for example, not oscillating. Loops that are stable can become unstable during test operation at different conditions, particularly as rates are lowered, and could require retuning during the test. Test participants should also recognize circumstances where it may be necessary to modify the existing control strategy or to operate with some control loops off-line to obtain the desired measurements. For example, determining the change of differential pressure as a function of throughput could be impossible if the column differential is on automatic control.

It is also important to prepare the testers for the method of data assembly and documentation required. All too often a great performance test is rendered useless because a protocol for proper documentation of the results was not established [21].

Another critical step is to have all instrument loops calibrated before the test begins. This will ensure that accurate data is collected when the test is conducted.

502.0 Test Procedure for Maximum Hydraulic Throughput

Many distillation columns will need to be operated at maximum throughput. Consequently, establishing the flooding point is important. Starting the procedure either at the normal operating point, or 50% of expected flood, or 50% of the troubleshooting performance point is recommended. In any of the capacity tests outlined in Section 306.4.1, the approximate capacity (or flooding point) can be found by increasing the feed rate and/or reboiler heat duty on the column in large steps at rapid intervals, allowing the column to line out before the next increment. As the flooding point is approached, smaller steps should be used to avoid premature flooding. The adjustment intervals may change with the tower sizes and loading levels, as well as sensitivity of downstream units to flooding conditions.

When the column shows symptoms of flooding, the load should be reduced until conditions become stable again.

The flood point should be determined after the column has reached thermal equilibrium, which may require 4-8 hours of operation after start-up. The monitoring of an overall heat balance with time can help determine when the column has reached thermal equilibrium.

502.1 Flood Symptoms

One definition of flooding is inoperability due to excessive retention of liquid inside the column. Flooding becomes apparent by one or all of the following, in order of decreasing reliability [22]:

502.1.1 The main telltale symptom of flooding is that the base liquid level cannot be maintained and keeps dropping because liquid cannot get down the column. The effects at true flooded conditions may manifest themselves in the following ways:

1. The level control valve regulating the bottoms product flow goes completely shut, while base level continues to drop.

2. Level in the reflux drum rises suddenly due to massive entrainment from the top tray, or liquid backing up the column from the flooded section.

502.1.2 Secondary symptoms at the onset of flooding are higher than normal pressure drop, followed by an increase in the temperature profile and a noticeable loss of separation efficiency.

1. Pressure drop increases markedly in the section where flooding takes place. Since only part of the column may be flooded, such as only the section above the feed point, the overall column pressure drop may not increase markedly. As a guide in trayed columns, one might expect the pressure drop of each flooded tray to approach the head of clear liquid up to one-half the tray spacing. Measuring pressure drop across each section of the column is recommended.

2. Column or section pressure drop continues to rise without any increase in vapor load.

3. Efficiency drops as the column approaches flood, but is technically still operable because hydraulically it can still pass liquid on a steady state basis. This drop in efficiency is caused by massive entrainment, leading to back mixing or a downcomer limitation backing up liquid for a few trays.

4. A rapid increase in temperature occurs below the flooded section because the reduced down flow of liquid from the flooded section leads to that liquid becoming abnormally stripped of its lower boiling components, and because the pressure drop has increased the boiling point of the liquid for a known composition.

5. An increase in reflux flow rate does not result in an increase in heat input required to maintain the same bottom column temperatures. The increased reflux first accumulates in the column, eventually leaves in the overhead, and never reaches

the bottom of the column. This is due to massive entrainment from the top tray.

6. A heat balance around the overhead condenser reveals liquid carryover into the vapor line from the column. This, of course, could be due to high entrainment and therefore is not conclusive of a flooded column.

7. The column will surge by pressuring up and then dumping. This type of cycle may occur before reaching the true flood point.

502.1.3 Other devices, such as vapor bleeders for high-pressure columns and sight glasses, give visual indication of flooded columns. A description of these devices is given in Kelley, Pickel, and Wilson [23], but modifications in their operation may have to be made for reasons of safety or pollution control. A radiation detector, such as a gamma-ray detector, can be very valuable. It can be used without any internal connection to show a high level of liquid or foam in a column or in the downcomer [23, 24].

502.2 Performing Capacity Tests

502.2.1 *Safety and Environment:* Test procedures must conform to all statutory and company safety and environmental regulations. The test plan should be reviewed with persons familiar with safety and environmental requirements and amended as necessary to comply fully.

502.2.2 *Capacity Test Strategy:* To find the flood point, either vapor or liquid flow rate or both are raised. Most commonly both are raised, because otherwise column material balance is affected and one product will have poor purity before flooding conditions are reached. The following techniques are commonly used for raising vapor and liquid rates during flood testing:

1. Raise feed rate, while simultaneously increasing the reflux and reboiler rates in proportion or in a way that keeps product compositions constant. This technique gives the most direct measurement of the maximum feed rate that can be processed through the column, but can only be applied when upstream and downstream units have sufficient capacity to handle the additional feed and products.

2. Raise the reflux and reboiler rates while keeping feed rate constant. This is probably the most popular technique used. Only two variables (instead of three) need to be changed, product compositions will not deteriorate, and the column will remain independent of the capacities available in other units, making it simpler and easier to implement. Usually, data

provided by this technique can be easily extrapolated to predict the maximum column feed rate.

3. Vary preheater or precooler duty while adjusting reflux and reboiler rates. This method, which can only be used if the feed is preheated or precooled, is often restricted by the exchanger capacity, and is least popular. In some multi-component distillations, it can give misleading results because it may induce accumulation of an intermediate impurity in one section of the column.

502.2.3 *Capacity Test Procedure:* Using any of the above techniques, reflux and reboiler rates will be varied. The procedure of varying these rates is important, and must consider the column control system.

Most column control schemes use the composition (or temperature) controller to manipulate either the reflux or reboiler, directly, or indirectly. The stream that is not controlled is commonly "free," i.e., is on flow control. This "free" stream is usually manipulated during flood testing, while the stream on temperature control is automatically adjusted to maintain product composition. For instance, if the reflux rate is on temperature control and the reboiler rate is on flow control, flood testing is performed by raising the reboiler rate. This causes the temperature controller to call for more reflux. Simultaneously, the additional boil-up raises column pressure and the pressure controller will increase condensation. The added condensation will supply the additional reflux requirement, and the column will reach new stable conditions with both reboiler and reflux rate increased. It should be noted that the increased reboiler and reflux rates will result in a changed column composition profile, and the temperature control set point may need to be adjusted to maintain the same product composition.

This procedure may induce the problem of "overshooting" the flood point. In the above example, reflux rate will increase shortly after a change in reboiler duty. The column may look stable for quite some time following the change, even if the reflux and reboiler rates which cause flooding have been exceeded, as it may take the liquid some time to fill the packing voids, tray decks, or downcomers to an extent sufficient to initiate flooding. This is particularly true for columns in which the ratio of liquid holdup to internal liquid flow is large. Meanwhile, vapor and liquid rates are raised further as the test progresses. When overshooting occurs, the flood point determined will be higher than the actual flood point. A stationary gamma source and detector can be placed on a tower at one or more strategic locations to detect an overshoot [25].

The problem of "undershooting" the flood point can also happen. Its occurrence depends on the dynamics of the column. For instance, in the above example, stepping up the reboiler rate too rapidly can cause a surge of additional liquid to be held up on the trays or in the packing. This extra liquid holdup may take up enough of the void space in the packing or the disengagement space between the trays to cause the column to flood prematurely. When this occurs, the flood point determined will be lower than the flood point at steady state.

To avoid overshooting or undershooting the flood point, little can be offered as a substitute to raising vapor (or liquid) rates in extremely small steps, and allowing long stabilization periods between steps. It may pay to carry out a preliminary flooding test, in which the steps are large and fast. Typically, vapor (or liquid) rates are raised by 5 to 10% increments at 15 to 30 minute intervals during the preliminary test. Increments as small as 1 to 2% are preferable, even in the preliminary test. It has been found [23] that frequent small increases in vapor (or liquid) rates are less likely to upset the column prematurely, and generally require shorter stabilization periods.

If available, dynamic computer simulations that do not assume steady state operation can be useful for providing guidance on appropriate step sizes to avoid disrupting column operation without excessively extending test duration.

Although the results of a preliminary test may suffer from overshooting or undershooting, they are likely to find the flood point within +/-10%, and often within +/-5%. The results of this test are used to find a good starting point for the final test. The preliminary test technique has been effective both for improving accuracy and reducing the time requirements for flood point determination [23].

During a capacity test within an operating plant, occasions exist when feed or auxiliary equipment capacity limitations prevent the column from reaching its true maximum capacity at flooding. Here, it may be necessary to over-reflux the column to introduce a maximum hydraulic limitation artificially, that is, temporarily pump more reflux to the column than is provided to the reflux drum by the condenser rundown. This necessarily results in the reflux drum liquid level being reduced and thus can only be maintained for a limited time. However, depending on the size of the reflux drum and the amount of over-refluxing required, it can probably be sustained for long enough to achieve flooding in a column where it would otherwise not be possible. Because of the unsteady state nature of this practice, it can only be used in capacity tests and is not appropriate for efficiency tests.

With large columns, a test lasting several days is sometimes best when accurate determination of the flood point is required. In most plant situations, weekend tests are ideal, as changes due to fluctuations in upstream units are minimized.

Accurate material and energy balances are important for flood point determination. These should close within 3 and 5% respectively and should be checked before the test and during the test. Generally, an accurate component balance is not needed. Several key guidelines described in Sections 503.3 to 503.6 are also useful for flood testing, particularly those concerning material and energy balances. However, flood tests are far less sensitive to analytical errors than efficiency tests, and therefore require a much lower level of effort.

The ambient weather conditions should be recorded during the flooding test. If possible, the flooding tests should not be carried out in extreme weather conditions that would encourage excessive heat loss, especially for columns with poor or absent insulation.

502.3 Optional Test Technique – Gamma Scanning [25]

An optional procedure to consider during testing is to use gamma scanning to monitor the hydraulic changes occurring and to pinpoint the origin of the flooding. Whether for packed or trayed columns gamma scans may provide a measurement of tray and distributor liquid levels, entrainment levels, and areas of liquid holdup.

For trayed columns, consider using gamma scans to study both the tray active areas for liquid level buildup and entrainment, and downcomers for liquid levels. For packed columns, a full scan (multiple scan lines) can be used if watching liquid distribution patterns is of interest, or a single scan line at each operating condition to measure liquid levels on internals and for areas of liquid holdup.

If gamma scanning is used, a baseline scan at the preliminary test conditions is suggested (see Section 503.3.6). Follow-up scans can be done at intermediate test conditions if interesting data are observed, but certainly at test conditions at or approaching flooding. Scans should be carried out while the column can be maintained at steady state conditions (Section 503.5).

503.0 Considerations Affecting Efficiency Test Procedure

503.1 Rigorous versus Shortcut Efficiency Tests

The procedure described below for efficiency testing is rigorous, demanding, and time consuming. Concerns are often expressed about the cost-effectiveness of the rigorous procedure, and it is often argued that a shortcut version should suffice. A suitable shortcut procedure can be derived from the rigorous procedure described here by skipping over guidelines considered less important.

The best procedure to adopt depends on the objective of the test. A shortcut test is best suited for detecting gross abnormalities, and is often performed as part of a troubleshooting effort. When investigating a gross malfunction, rigorous testing is seldom justified because identification and correction of the fault will likely be delayed. When a column appears to perform well, a shortcut test can provide a useful, although somewhat unreliable, set of data for future reference.

A rigorous test is best suited for detecting subtle abnormalities, for accurately determining column efficiency, for checking the design, for optimization, and as a basis for performing improvement or debottlenecking modifications. Shortcut tests applied for these purposes often need repeating, yield conflicting data, provide inconclusive results, or lead to ineffective modifications.

Shortcut tests that do little more than take a set of readings and samples from the column are best avoided. Shortcut tests should, at the minimum, incorporate checks of material, component, and energy balances and some key instrumentation. These checks will permit identification of major problem areas and a rough assessment of data reliability. These key items can be extracted from the list of preparations and checks recommended below for rigorous tests.

503.2 Strategy of Efficiency Testing

503.2.1 *Safety and Environment:* Test procedures must conform to all statutory and company safety and environmental regulations. The test plan should be reviewed with persons familiar with safety and environmental requirements and amended as necessary to comply fully.

503.2.2 *Scope:* In a multi-column unit, it is best to carry out a performance test covering the whole unit. Testing these columns one at a time increases the total effort and time consumed, and reduces the reliability of measurements. Testing the entire unit provides several material balance cross-checks and permits better identification of incorrect meters and lab analyses. For instance, if the column feed analysis is off, the column component balance may not be sufficient to point out which analysis is suspect; but if data from a component balance on upstream and downstream equipment are also available, the incorrect analysis can be identified.

A shortcut for the above recommendations may be acceptable when the plant operating rate and compositions have not significantly changed since the last plant test and the problem areas are well known. In such cases, it may suffice to test only the specific column of interest.

Another instance when a shortcut is often acceptable is when the column is near the end of a processing train that yields reasonably pure products. Product analyses and metering are far more reliable than intermediate stream measurements. Usually more can be gained from cross-checks with downstream measurements than with upstream measurements. In such cases, it may suffice to restrict the testing to the column area and downstream equipment.

503.2.3 *Duration:* Carrying out an efficiency test over a two- to three-day period is best. If shorter periods are used, variations in plant conditions may introduce serious errors. Over a period of two days, errors are averaged out. Furthermore, column control problems may make it difficult to obtain a sufficiently long period of stable operation if the test is short. Over a two-day period, the column should be running under stable conditions for a significant duration.

If product or charge tanks are to be gauged to obtain flow rates, the test period should be long enough to measure tank displacements within 1 or 2%.

503.2.4 *Timing:* The best time to carry out an efficiency test is when the plant is stable. In most plant situations, weekends are ideal, as changes due to fluctuations in upstream units are minimized.

503.2.5 *Personnel:* When the data collection effort is extensive, assembling a special data gathering team with a team leader is best. The team leader is responsible for assigning specific responsibilities, keeping work schedules, and coordinating test preparations and

503.3 Early Preparation for Efficiency Tests

Preparation is the most important phase of an efficiency test. A malfunctioning meter, a leaking block valve, or a poor laboratory analysis during the test can dramatically reduce the reliability of the results and defeat the purpose of the test. This is the time to identify all potential problems and complications [7, 26-29].

503.3.1 *Preparation Checklist:* This should include a detailed list of instruments to be tested and required laboratory analyses. The instrument list should ensure that no important meters are omitted and that those included are sufficient to entirely define and check unit performance. Temporary instrumentation should be added to fill any gap in the information. Any handheld monitoring devices, such as timers or contact pyrometers, should also be included in the list.

All meters need to be surveyed to ensure that they are operational, their calibrations are valid, their indicators are reading on-scale, and that there are no oddities in field instrument installation. For instance, a flow meter will read inaccurately when an orifice plate is incorrectly sized or damaged, or when a tap is plugged. All orifice plates must be checked to ensure they are the proper size and in good condition. Failure to check orifice plates has ruined the results of many test runs. All orifice taps should be blown, and pressure indicator taps should be checked for any blockage. All dial thermometers and pressure gauges should be checked for accuracy, and unsatisfactory ones should be replaced or removed before the test. A thermocouple or a thermowell may read incorrectly if it is too short or coated with deposits. Attention should be paid to impulse lines, heat tracing equipment, and vibrations near transmitters. Any oddities should be corrected before the test.

The location of sample points to be used in the test should be determined. Any cooling coil requirements should be identified. All sampling lines need to be operable, free of any blockage, and comply with the relevant safety requirements.

Testers must also compile a list of all samples required for the test and their frequency. The list should identify any key samples that need to be taken in duplicate for verification of results. Any samples that need special handling, such as refrigeration or stabilization, should be individually noted, and the sample containers suitable for each analysis—sample vessels, glass bottles, plastic bottles—and the best way of sealing the containers, such as plugs, corks, or screwed tops, should be selected.

At this stage, it is important that the efficiency test team check with the lab, safety, production, and environmental personnel to determine what measures and equipment, such as personal protective equipment, are needed for sampling and other test duties. Any unavailable items should be marked on the checklist and ordered. The list should clearly identify the type and quantity of items required and who is responsible for ordering them.

If gamma scanning will be used to map the hydraulic conditions at flooding, getting a baseline scan at the preliminary test conditions is suggested.

503.3.2 *Test Date:* Ideally, the test should be set at a time most convenient to the plant production people. Stable operation during the test is essential, and the production personnel can anticipate periods of potential instability. Another major consideration is the availability of laboratory personnel, since an efficiency test usually means increasing the lab workload. The longer a sample sits on the shelf, the lower the reliability of the analysis because of possible leakage or chemical reaction, and the greater the chance of the sample being lost.

503.3.3 *Process Control System and Data Acquisition:* The reliability of control and data acquisition systems and the extent to which they can be used without interfering with normal plant operation needs to be checked. Acquired data should be compared to control board readings and any discrepancies investigated. When more than one system is used to acquire data, such as a distributed control system and a separate on-line process analytical system, all clocks should be synchronized and any offsets noted so that data obtained on the different systems can be matched to each other.

Lastly, the efficiency test team should explore the option of having a control computer or data acquisition system programmed to prepare a special report of data, to store key data points more frequently, to trend key variables, and to add normally unmonitored data to its records.

503.3.4 *Computer & Data Processing:* Since material and component balances and efficiency calculations are often carried out with computer tools, such as process simulation programs and spreadsheets, preparing those tools and applications before the tests may prove very beneficial. Once the tools are available, process and analytical data can be quickly processed and inconsistencies easily detected before or during the test.

503.3.5 *Data Sheets:* It is best to use modified versions of the log sheets or data entry forms normally employed by the shift personnel to enter data. This is particularly important if the shift team is requested to take readings over the night shift. A totally new data sheet may generate confusion and reduce cooperation.

Readings such as levels and control valve positions, and the times at which readings are taken may appear unimportant, but can often turn out to be valuable in analyzing test results. For instance, a reboiler control valve position can provide useful information for determining if the steam side is flooded with condensate. These should be included on the data collection sheets.

503.3.6 *Preliminary Test:* The preliminary test allows validation of the items on the data sheet and identifies any measurements that may have been omitted. Finding these omissions during the test is too late. It also provides an early check of the closure of material and energy balances. Leaking valves and exchanger tubes affect the material balance, and these leaks need to be identified before the final test. Meters may read incorrectly, even at times when instrument technicians are sure they are right. A material and component balance closure within $\pm 5\%$ and an energy balance closure within $\pm 5\%$ are recommended for the test (See Sections 602.0 and 603.1).

Attention needs to be paid to flow measurements of streams close to their bubble point or dew point. Flashing of liquid or condensation of vapor can lead to major errors. Any "noisy" or heavily dampened flow meter signals may suggest unexpected flashing or condensation. Whenever practical, it pays to cross-check the material balance against direct volume or weight measurements, such as those from product tank levels.

Component balances can often be checked by plotting the ratio of distillate to bottom concentrations for each component against its relative volatility based on a reference component, such as the heavy key component, at average column conditions. Except for columns that are primarily rectifying or stripping, a log-log plot should give a straight line or a smooth curve approximating a straight line.

Doing an energy balance is important. An incorrect reflux meter will not show up in the overall material balance since it is an internal stream. The ambient weather conditions should be recorded during the tests.

Attention should be paid to exchanger duties calculated from small temperature differences, such as a condenser duty calculated from the inlet and outlet temperatures and cooling water flow, where the temperatures are less than 10 °F (6 °C) apart. Often the flow can be throttled to increase the temperature difference; if this is impractical, high-accuracy temperature indicators may be required.

503.3.7 *Monitoring the reliability of laboratory analyses:* Lab analyses should be checked against design data, previous test results, or component balances, and any inconsistencies investigated [6,28].

503.3.8 *Preparing a worksheet for the unit:* This is a data table showing items of equipment, stream names, tag numbers, temperatures, pressures, flows, and key component compositions. This table is used to take key readings during the test and to troubleshoot for gross deviations. The worksheet can be done as an electronic spreadsheet or a paper data sheet. It is helpful to conduct a "dry run" using this sheet before the test. See Table 2 for an example of a data table for a heat pumped C_3 Splitter.

503.3.9 *Checking filters, coalescers, and storage tank levels:* Efficient operation of filters, coalescers, and storage tanks are particularly critical to a successful performance test. If sufficient operating margin is not available in even one such unit, it may interrupt or disturb the test. The status and condition of these types of units should be checked to see that they are adequate before the test.

503.4 Last-minute Preparations for Efficiency Tests

Final preparations are critical for a successful efficiency test. Last minute problems are most likely to occur during the test and may lead to meaningless test results [6, 7, 26-29]. Check to see that:

503.4.1 All test supplies and labeled sample containers are on hand and required safety equipment has been received by those intended to use it. Hand-held instruments are rechecked.

503.4.2 Shift supervisors and laboratory personnel are aware of the test and of final safety requirements.

503.4.3 Instruments and sample points are tagged and labeled with weatherproof tags. This reduces the possibility of reading the wrong instrument or taking the wrong sample.

Tower Name :	C3 Splitter				
Item Number:	A-89				
Date:				3/2/2011	3/3/2011
Time:				9:00 AM	9:30 AM
Description	**Measurement**	**Tag No.**	**Units**	**Value**	**Value**
Feed	Liquid Flow	FIC5028	MPH	390	389
	Temperature	TI-5311	°F	138	138
	Pressure	PI-5337	PSIG	304.7	305.1
	C3H6 Composition	AI-9037-1	Mole %	30.44	34.11
	C3H6 Composition	Lab	Wgt%	29.49	33.67
Tower	Top Temperature	TI-3763	°F	55.74	55.2
	Top Pressure	PI-6017	PSIG	107.31	107.6
	Bottom Pressure	PI-6004	PSIG	122.81	122.91
	Upper Delta-P	PDI-6024	PSI	11.2	11.14
	Lower Delta-P	PDI-6004	PSI	4.2	4.18
Reflux	Drum Pressure	PIC-6049	PSIG	187.52	187.56
	Temperature	TI-0397	°F	59.7	60.96
	Flow	FIC-6023	MPH	2428	2430
Propylene Product	Flow	FIC-6055	MPH	105	117.1
	Temp	TI-6053	°F	87.4	87.1
	C3H8 Composition	AI-9045	Mole ppm	25324	22623
	C3H8 Composition	Lab	Wgt, ppm	26770	22340
Propane Product	Flow	FIC-6070	MPH	268	265
	Temp	TI-6069	°F	76.56	76.74
	C3H6 Composition	AI9037-2	Mole %	2.85	3.54
	C3H6 Composition	Lab	wgt%	2.66	3.21
Material Balance	Overall Balance			4.4%	1.8%
Material Balance	Propylene Balance			5.0%	6.1%

Table 2. Example of Data Acquisition Table for Heat-Pumped C$_3$ Splitter

503.4.4 Sufficient blank data sheets are available with weatherproof sheets to cover them while they are carried around the plant. Difficulties encountered during data recording may lower the quality of the data and the cooperation of those taking readings. Any ideas that make data recording easier are highly recommended [6, 27].

503.4.5 Safety valve bypasses, steam traps, and drains are checked for leakage.

503.4.6 Refrigerated heat exchangers are de-oiled as necessary. Consider injecting methanol into columns susceptible to hydrate formation. Strainers should be back-flushed.

503.4.7 A sufficient safety margin is allowed between test operating conditions and maximum process limits. A popping relief valve will disrupt a test.

503.4.8 The test is scheduled when no unstable weather conditions are expected. Plant personnel will be too busy keeping the plant on-line if instability is likely and the weather may adversely affect instruments. An unexpected heavy thunderstorm can severely hamper data collection and worsen unit heat loss.

503.5 Establishment of Steady State Conditions

Steady state must be reached at each test condition before the beginning of the test. It must be maintained throughout the entire run. This state is recognized by observing when column temperatures, pressures, levels, and flow rates have remained constant for a period. For columns in which high purity products are made, the temperatures in the section that has the highest temperature gradient are the most sensitive and are the best criteria for steady state conditions. The time required to reach steady state will be greater for columns containing a large holdup in relation to the throughput. This time should be sufficient to displace the liquid holdup in any part of the column and auxiliaries, including reflux drum and surge in the reboiler circuit, with liquid overflow from two to ten times, depending upon relative volatility and the desired fractional approach to equilibrium [30]. Some special considerations are:

503.5.1 In cases in which samples can be analyzed rapidly or continuously, the steady state condition may be determined by sampling the overhead, bottoms, and feed. At least three successive samples should show essentially constant composition. For columns in which high purity products are made, an additional sampling point where the concentration gradient is greatest is useful.

503.5.2 Factors that may prevent reaching steady state are listed in Section 706.6. These factors should be reviewed before and during the test so that their impact can be reduced.

503.6 The Test Day

If preparations have been adequate, the test should go smoothly. The main considerations in this period are:

503.6.1 Test leaders should ensure that all personnel wear protective equipment and follow safety procedures. Unsafe practices are not only hazardous, but can also generate friction with operating supervisors and may lead to discontinuation of the test.

503.6.2 For older plants, any strip charts should be marked with time notes. Time markings should be made at the beginning and the conclusion of the tests and at several times during the test. Operators may "move charts along" when inking recorder pens.

503.6.3 The worksheet referred to in Section 503.3.8 should be continuously watched and major inconsistencies immediately explored.

503.6.4 Close communication with the plant superintendent, shift supervisor, panel operator, and upstream and downstream unit control rooms are essential. Miscommunication often leads to plant decisions that unnecessarily lower data reliability.

503.6.5 The sampling technique should be watched when key samples are taken. A common problem is not allowing sufficient purge time through the sampling line and sample vessel, especially when sample lines are long. For key samples, one useful technique is to obtain these in duplicate and keep one aside in case the other is lost.

503.6.6 Taking readings around the clock is best. If test duties are delegated to the shift team overnight, it pays to check if they know exactly what is needed. A phone call to the plant may help sort out problems and will affirm the importance of the test to night shift personnel. In some cases, it may be necessary to be present with the overnight shift people during the test simply to show its importance.

Recognizing operating personnel with food items, such as doughnuts in the morning or ice cream at night, may be appropriate. Simple displays of appreciation like this can be very effective at smoothing out a lengthy plant test.

503.7 Concluding Test

No test is complete until all special equipment, such as instruments and labels, are properly removed and the plant is returned to its initial status. Any temporary equipment remaining may become a safety hazard or a nuisance; any temporary instruments may become damaged or lost.

Labels may confuse the operators. Poor tidy-up is one of the most frequent sources of complaints against test organizers.

It is best to first obtain input from plant supervisors about which items need removing— and how soon—and which items, such as some labeling, may be a beneficial addition to the plant. Any items to be retained should be discussed with the department supplying them. The rest should be immediately removed.

> **503.7.1** *Follow-Up:* A letter or memorandum to the unit operations staff acknowledging their cooperation and reinforcing the importance of the test will be greatly appreciated and will promote good will.

600.0 COMPUTATION OF RESULTS

601.0 Verification of Test Data and Simulation Models

Both the test data and simulation models must be verified as reliable before capacity or efficiency calculations can be made. Capacity calculations require constant flow rates and temperatures, but changes in composition that do not affect the enthalpy balance by more than 5% are allowable. However, efficiency calculations are more sensitive to compositions, so steady state conditions are required. Vapor-liquid equilibrium (VLE) models used for stage-to-stage simulations must be verified with VLE data or experience [31]. The test data should be used to verify any computer models developed for the process.

602.0 Material Balance

The overall material and individual component balances must be made around the column. The allowable error depends on the test objectives, but, ideally, the overall balance for the column should agree within ±5%. Ideally, the balances on the individual distributed components should also agree within ±5%, and may need to be tighter for careful efficiency determinations. The allowable error in the balances of the minor components will depend partially on the concentrations of those components. For constant absolute analytical precision, the lower the concentration of a component, the greater the percentage of error in its balance is likely to be. The accuracy with which balances can be made also depends on the test system. For example, with pseudo-components in petroleum fractions or dilute components, ±10% may be more realistic.

The percentage of error for a minor component material balance will inherently be greater than the overall material balance. Obtaining a minor component material balance within the allowable error is often difficult. In addition, if any of the minor components can form a binary azeotrope with one of the key components, or form a multi-component azeotrope, the target product purities, based on the feed composition, may be unachievable.

When a high degree of accuracy is required for the minor components, determining the component amount is sometimes easier based on downstream column product sampling data, where such minor components have a higher concentration.

If these conditions are not met, those people performing the test should understand why and be careful in interpreting the results.

If the balances are unsatisfactory, the data should be examined to make sure that the proper corrections to the instrument readings have been made. The compositions may be checked by means of equilibrium data and the observed temperatures and pressures. An empirical check on the concentrations of the distributed components in the overhead and bottoms streams may be made by plotting, for each component, the ratio of its concentration in the overhead to that in the bottoms versus its relative volatility at average column conditions. Except for columns that are primarily stripping or rectifying, or where relative volatility varies greatly in the column, the plot on log-log paper should give a straight line, or a smooth curve approximating a straight line.

The measured flows and compositions should be adjusted to balance the mass flows around the column before any capacity or efficiency calculations are made. Sometimes, the flows can be adjusted to improve the component balances. Adjustment of the flows by a least squares method, as described in Gelus, Marple, Jr., and Manning, Jr. [32], should be considered if many distributed components exist. Refer to Section 701.1.1 for a list of typical causes of poor material balance closure.

602.1 End Effects

When determining packed bed or tray efficiency from reflux, feed and bottom samples, allowances must be made for the separating effect of the reboiler and overhead systems. A kettle reboiler has been found to act as one theoretical stage, but a thermosyphon or falling film reboiler and a partial condenser may provide less or more than one stage, depending on design and takeoff ratio. A distributor or chimney tray may act as a partial stage when it is mounted so as to expose falling liquid to the vapor stream. Proper handling of end effects becomes less important as the tower height or stage number increases.

603.0 Enthalpy Balance

603.1 Overall Balance

Before the flows inside the column can be accurately calculated, the consistency of the thermal data for the external streams must be checked. The measured energy input should be taken as the net enthalpy contribution of the entering process streams and the energy added to the reboiler or inter-reboiler (if used). The energy output should be taken as the enthalpy of the leaving process streams, with the heat removed by the condenser and any circulating reflux loops (i. e. pumparounds). Often, losses to the surroundings, due to radiation, convection, or conduction, are significant and need to be estimated and taken into account. The same standard state temperature data should be used to calculate the enthalpy of the entering and leaving process streams.

Ideally, the thermal data should be sufficiently accurate such that the calculated energy input and output agree to within ±5%. Realistically, +/- 10% enthalpy balance is much more achievable. If not, those people performing the test should understand why and be careful in interpreting the results.

An example of an enthalpy envelope for an entire system is represented in Figure 6 by the solid line labeled "Overall Balance." Several other bases for the enthalpy balance can be used. These are shown by the dashed lines labeled Alternate 1 and Alternate 2 in Figure 6. Consideration should be given to the possibility that some process streams passing through the envelope may contain both vapor and liquid phases. If both phases are present, accurate information on their proportions and compositions is required. Otherwise the envelope may need to be adjusted such that all streams entering and leaving are single phase. This information can be used to help in determining which envelope should be used for the enthalpy balance. Having the data for making all three alternative balances is desirable. Then a balance around the feed preheater can be used to check the condition of the feed, and a balance around the condenser can be used to determine if overhead entrainment is present.

When the distillate is not one pure component, an alternative method for determining overhead entrainment is available. The presence of large quantities of entrainment is indicated when the overhead temperature is lower than that calculated from the composition of the overhead vapor or when the condensate flow is higher than can be accounted for by the heat balance. Refer to Section 701.1.2 for a list of typical causes of poor closure of enthalpy balances.

603.2 Internal Flow Rates

Internal flows are determined by making local heat balances (Figure 6) or from computer simulations of the column. For maximum (or minimum) capacity tests, the material balance envelope should intersect the column at the plane where the internal volumetric flows are the greatest (or least). Preliminary balances may be needed to determine the location of the maximum (or minimum) loading. The determination of operating lines for the calculation of efficiencies requires making local balances at several locations in the column where major changes occur in the internal flows. If internal samples are not available, the compositions can be estimated with enough accuracy for enthalpy balance purposes. Hengstebeck [3] provides a method for estimating intermediate compositions in an operating column with a multi-component system.

The task of determining the internal vapor and liquid flows and compositions is much easier if a verified distillation simulation model is available on a computer [33]. The model can be used to calculate internal flows and external heat duties given the feed, distillate, and reflux rates shown in Figure 6. Other combinations of feed, distillate, bottoms, and reflux can be used if certain data are more reliable.

604.0 Hydraulic Performance

604.1 Trayed Column

The internal flows for a column are calculated from the local heat balances or are available from the column computer calculation (Section 603.2). These flows can be used to determine the maximum (or minimum) throughput for an acceptable separation and/or the maximum hydraulic load for the column. The column may reach an operating poin, before the hydraulic limit where acceptable specifications cannot be met on the process streams. The internal vapor and liquid rates can be used to explain the loss of separation efficiency at these rates.

The internal flows for the column are also useful for calculating the pressure drop to be expected during normal operation (Section 802.3). The comparison of the calculated to the actual pressure drop is very useful in determining if the column is operating properly and how close the system is to the flood point. Many published tray pressure drop correlations do not incorporate the static head of vapor. This omission is usually significant in high-pressure columns (Refer to Section 404.2.6).

The maximum hydraulic limit occurs when any one tray in the column reaches its capacity limit for handling the vapor or liquid traffic in the column. The flooding of a single tray will cause several (or all) trays above that point to fill with liquid. The location of this tray may be evident from observations, as described in Section 502.1, or by gamma scanning, as explained in Section 502.3. The vapor and liquid flows for this point should be calculated. If the test run did not define this location, calculations should be made to determine the maximum load point. This point is most easily determined by reviewing the vapor and liquid loading profiles predicted from a computer simulation.

The calculated maximum flows should be used with published correlations to determine if the column is flooding prematurely. Several correlations are available for various tray types, including bubble cap trays [34-36], and sieve and valve trays [37-45]. FRI and major vendors also offer computer programs for hydraulic ratings of various tray types. If a comparison with these correlations shows that the column is flooding prematurely, additional factors that might affect performance—tray damage, plugging, improper installation, and foaming—should be considered. Other possible causes for premature flooding are discussed in Section 704.0.

604.2 Packed Column

The capacity of a packed column depends upon both the packing and column internals. The column internals must suit the packing and must not limit the capacity of the column. Components of a typical packed column are shown in Figure 1. Refer to Kister [6, 46] and Strigle [47] for more details on packings, distributors and other internals.

The internal flows and maximum hydraulic load for packed columns can be determined in the same way as trayed columns. The maximum hydraulic limit occurs when any section in the packed column reaches its capacity limit for handling the vapor or liquid traffic.

The flooding at a section of packed bed, say at the bottom of the bed, will cause part or all of the bed above that point to fill with liquid. The location of this flooded section may be evident from observations as described in Section 502.1 or more directly by gamma scanning in Section 502.3. The vapor and liquid volumetric flows for this point should be calculated. If the test run did not define this location, calculations should be made to determine the maximum load point by a computer simulation.

Compared with trayed columns, the initial flooding point in a packed column may not be easily recognized by pressure drops or gamma scans because of the continuous nature of the liquid and vapor flows in the packed beds and the relatively high bed metal densities. This would be especially true for low liquid load applications. A baseline column scan during normal operations is recommended (Section 502.3).

The calculated maximum flows should be used with verified correlations to determine if the column is flooding prematurely. Several correlations available for various packings are summarized in Kister and Strigle [46, 47]. FRI and major vendors also offer computer programs for hydraulic ratings of various packings. If a comparison with these correlations shows that the column is flooding prematurely, additional factors that might affect performance should be considered, including damage to the packing or distributors, plugging/fouling, improper installation, and foaming. Other possible causes for premature flooding are discussed in Section 704.0.

Pressure drop of packing can be estimated with the modified generalized pressure drop correlation (GPDC) by Strigle [47] (See Table 3). Typical packing factors in use with the modified GPDC are given in Tables 3 and 4. Sample calculations using the GPDC method are shown in Section 802. For a given proprietary packing, manufacturer's correlations are best used.

Log-log plots of pressure drop vs. gas rate are usually linear at low to moderate gas rates, with liquid rate as a parameter (See Table 4 and Figure 15). At higher gas rates, however, the wet pressure increases more steeply because the liquid holdup increases significantly with gas rate. Where this occurs, operation is said to be in the "loading zone." Eventually the pressure drop curve becomes nearly vertical at the flooding point of the packing.

605.0 Efficiency Performance

The overall column efficiency is useful for comparing the results of one test with those of another, or for comparing test results with the design expectations. For these comparisons to be valid, testers must use the same method of calculation, and the same equilibrium and enthalpy models.

605.1 Trayed Columns

To obtain maximum information from the test data, an equilibrium-stage simulation of the column is recommended. The calculation assures that the equilibrium conditions and heat and mass balances are satisfied for each theoretical stage. Before any calculations are attempted, it is essential that the consistency of the composition data be checked, as described in Section 602.0. Most computer programs require the number of theoretical stages as an input variable. The usual procedure is to make computer runs with various numbers of theoretical stages in different sections of the column to match the field test data. A section is a part of the column over which the vapor and liquid rates are relatively constant, and are typically bounded by the top and bottom trays, any other intermediate feed or draw trays, or locations of intermediate heat addition or removal. For example, a column with two feeds and one draw has four sections.

The test conditions are used to define the rest of the input data. These runs will determine the number of theoretical stages in each section needed to yield the product and draw stream compositions that most closely agree with the test results. The efficiency for each section in the column is probably different, so overall section efficiency is defined by the following equation:

$$Efficiency,\% \text{ (section)} = \frac{number\ of\ theoretical\ stages}{number\ of\ actual\ trays} \times 100\% \quad \textbf{(605.1.1)}$$

Packing	Size, mm or #	Packing Area, ft²/ft³	Packing Area, m²/m³	Void Fraction, %	Packing Factor, ft⁻¹	Packing Factor, m⁻¹
Metal Pall Ring	16	110	360	92	78	256
	25	62	205	94	56	183
	38	40	130	95	40	131
	50	32	105	96	27	89
	90	20	66	97	18	59
IMTP (I-Rings)	25	63	207	97	41	134
	40	46	151	97	24	79
	50	30	98	98	18	59
	70	18	60	98	12	39
Nutter Rings	#0.7	69	226	98	39	128
	#1	51	168	98	30	98
	#1.5	38	124	98	24	79
	#2	29	96	98	18	59
	#2.5	25	83	98	16	52
	#3	20	66	98	13	43
Raschig Super-ring	#0.3	96	315	96	52	171
	#0.5	76	250	98	42	138
	#0.6	66	215	98	37	120
	#0.7	53	175	98	30	100
	#1	49	160	98	25	82
	#1.5	35	115	98	18	59
	#2	30	98	98	15	49
	#3	24	80	98	11	36
Metal Jaeger Tri-Packs	#1	36	118	96	26	85
	#2	23	75	98	14	46
Hiflow	25	91	298	96	-	-
	50	53	175	98	16	52
Metal Raschig Rings	25	56	185	86	144	472
	50	29	95	92	57	187
	75	20	66	95	32	105
Ceramic Raschig Rings	25	58	190	74	179	587
	50	28	92	74	65	213
	75	19	62	75	37	121

Table 3. Packing Factors for Random Packings (Typical) [1]

Packing	Size or number	Packing Area, ft^2/ft^3	Packing Area, m^2/m^3	Void Fraction, %	Packing Factor, ft^{-1}	Packing Factor, m^{-1}
Mellapak	2Y	68	223	99	14	46
	250Y	76	250	98	20	66
	350Y	107	350	98	23	75
	500Y	152	500	98	24	112
	2X	68	223	99	7.0	23
	250X	76	250	98	7.9	26
Mellapak Plus	252Y	76	250	98	12	39
	452Y	107	350	98	21	69
Flexipac	2Y	67	220	99	15	49
	1.6Y	88	290	98	18	59
	1Y	128	420	98	30	98
	250Y	76	250	99	-	-
	2X	67	220	99	7.0	23
	1.6X	88	290	98	10	33
	1X	128	420	98	16	52
Flexipac High Capacity	2Y HC	67	220	99	13	43
	1.6Y HC	88	290	99	17	56
	1Y	128	420	98	25	82
Intalox	3T	52	170	99	13	43
	2T	66	215	99	17	56
	1T	94	310	98	20	66
Raschig Super-Pak	250	76	250	98	12	39
	350	107	350	98	21	70
Ralu-Pak	250YC	76	250	98	20	66
Rhombopac	6M	107	351	-	18	59
Montz-Pak	B1-250	76	250	95	20	66
	B1-250M	76	250	95	13	43

Table 4. Packing Factors for Typical Structured Packings [1]

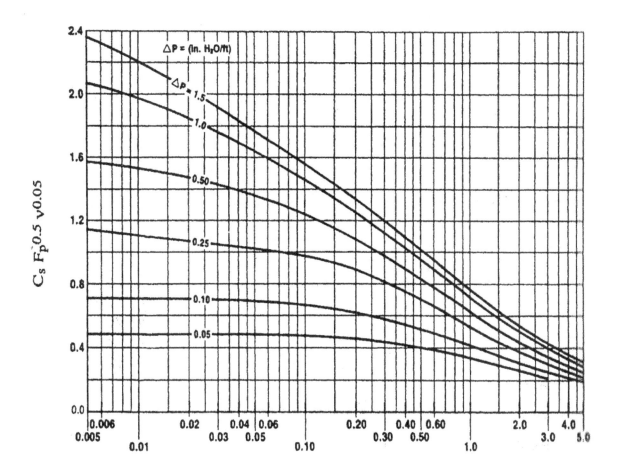

$$\frac{L}{G}\left(\frac{\rho_V}{\rho_L}\right)^{0.5}$$

**Figure 13. Generalized Pressure Drop Correlation of Eckert
as Modified by Strigle [1]**

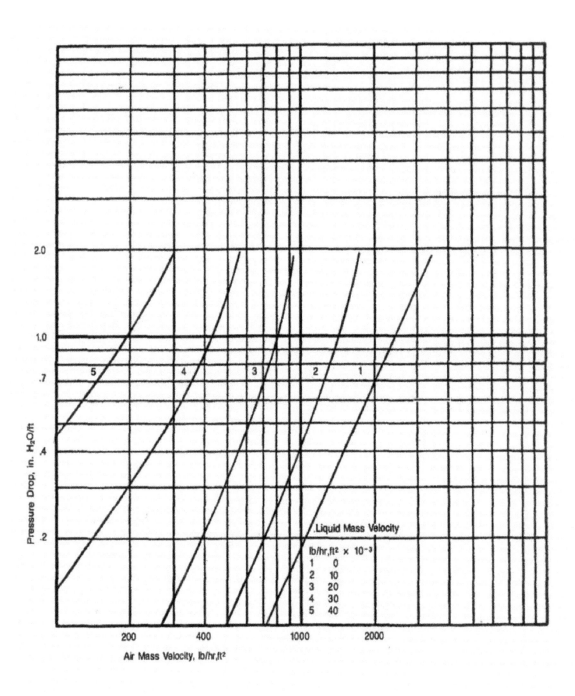

**Figure 14. Pressure Drop Curves for Random Packing
(Typical Manufacturer's Data for 1" Plastic Slotted Rings)**

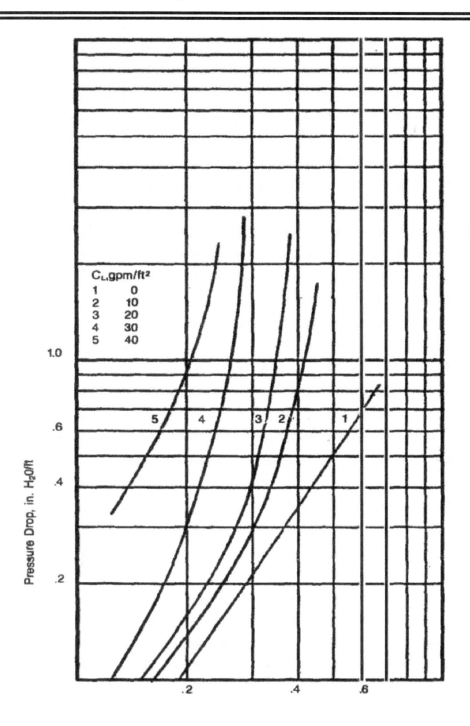

**Figure 15. Pressure Drop Curves for Structured Packing
(Typical Manufacturer's Data for No. 2 Corrugated Sheet Metal Packing)**

The overall column efficiency can be calculated from the same equation by substituting the total number of stages of the column in the equation.

Interpreting the results of computer stage-to-stage calculations may be difficult, particularly if no obvious key components exist. A method of finding the column efficiency in such cases is to plot, for each component of the feed, the computer-predicted percentage recovery of that component in the overhead as ordinate vs. the number of theoretical stages as abscissa. The actual percentage recovery in the overhead is then marked on the curve. The corresponding abscissa is the number of theoretical stages of separation given by the column for that component. The indicated numbers of theoretical stages for the different components will not exactly agree. This is due to inaccuracies in the analysis or the physical data, or because of differences in the mass transfer rates of the different components. The overall column efficiency for any component can be determined by using this plot and the equation presented above. For example, in Figure 16, the stabilizer has 45 actual trays. The recovery of i-Pentane overhead was about 11%, and the recovery of 1- Butene overhead was

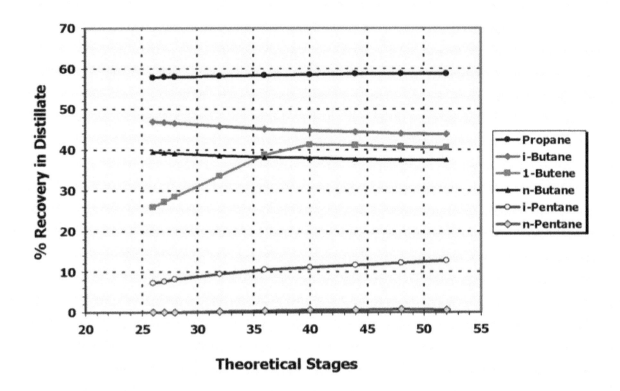

Figure 16. FCC Stabilizer Component Recovery Plot

about 38%. The resulting efficiencies of the trays based on these two components are 88% for i-Pentane (40/45) and 80% for 1-Butene (36/45).

Computer simulations based on a rate-based non-equilibrium model can predict vapor and liquid compositions on actual trays. If the underlying mass transfer models are accurate, the tray efficiencies for different components can then be back-calculated from the compositions on each of the actual trays [3].

605.1.1 *Murphree Tray Efficiency:* Apparent Murphree tray efficiencies can be calculated only when tray samples have been obtained at appropriate points. The efficiencies vary throughout the column, changing with the ratio of the slopes of the equilibrium and operating lines, and with composition and velocities in the column [34]. Since Murphree tray efficiencies are determined at specific velocity, composition, and slope ratio, they are valuable for correlating with fundamental data.

If samples of both the liquid and the vapor to and from the same tray are available, substitution in the formula for the Murphree tray efficiency [34, p. 7] is straightforward for binary systems. Usually liquid samples are only available every five to ten trays. A computer calculation can be matched against the data for that section, or the graphical method of Baker and Stockhardt [48] can be used to obtain the efficiency. Baker and Stockhardt's method requires the use of key components for multicomponent systems. An example of this application is given in Gerster, Mizushina, Marks, and Catanach [49]. Taylor, Davis, and Holland [50] also details the determination of tray efficiencies from operational data.

605.2 Packed Columns

The height equivalent to one theoretical plate (HETP) and the height of a transfer unit (HTU) are the two conventional models for packed tower mass transfer efficiency analysis. These are equilibrium-stage and rate-based methods, respectively. In the equilibrium- stage model, the bulk streams leaving a "contact stage" are assumed to be in equilibrium with each other. In the rate-based model, the equilibrium condition is assumed only at the vapor-liquid interface, while the compositions and temperatures of the bulk vapor and the liquid streams are handled by the respective rate equations. Usually, the theories discussed in this section are not applicable to processes with chemical reactions.

605.2.1 *The HETP Model* assumes that a packed bed can be treated as a series of theoretical plates or stages. Vapor and liquid streams leaving each stage are in equilibrium. With this assumption, theories developed for stage-wise mass transfer can be applied to the performance of packed towers. In reality, though, a packed tower is a continuous differential contactor [51].

The exact form of the equation for the HETP model is:

$$h_T = \sum_{i=1}^{N_{TP}} (HETP)_i$$

Where h_T is the total depth of packing bed and $(HETP)_i$ represents the HETP values for the i-th theoretical plate. However, because this equation is not readily applicable for design or data analysis, a second major assumption is introduced to the HETP model: that the HETP values for all the theoretical stages in a given section are the same.

With this assumption, Eq. 605.2.1.1 can be simplified to:

(605.2.1.2)

$$h_T = (HETP)(N_{TP})$$

Where N_{TP} is the total number of theoretical plates.

Equation 605.2.1.1 has been the prevalent form of the HETP model ever since it was first introduced by Peters in 1922 [52].

605.2.1.1 HETP Analysis Procedure

The procedure for packed tower efficiency analysis is as follows:

1. Following the recommendations given in Sections 601, 602, and 603, prepare all the measured data, which would include temperatures, pressures, flow rates and compositions of all the input and output streams, plus any additional temperature or composition data collected along the packed bed section bounded by these input/output streams.

2. Calculate the N_{TP} accomplished by the packed bed at the test condition. This can be accomplished with a validated computer simulation. The calculated composition of all streams, and the temperature and composition profiles along the bed and internal flows should agree with the measured values within an acceptable tolerance. On a fundamental level, the key component compositions of the product streams should match the simulated results.

3. If reasonable agreement is achieved in step (2), the HETP of the packed bed can be calculated by Eq. 605.2.1.2. (See the sample calculation in Section 802.2).

605.2.1.2 Methods of N_{TP} Calculation: For an understanding on the principles of N_{TP}, as illustrated by the McCabe-Thiele method for binary systems [53], materials in Wisniak, et al. [54] and King [55] are highly recommended. Both King [55] and Thompson [56] can be helpful for multi-component systems, while for the rigorous equilibrium-stage method of N_{TP} calculation, which takes into account both material balance and energy balance, the works of Holland [57,58] should be consulted. Applications to complex petroleum mixtures are discussed in-depth in Van Winkle [59] and Watkins [60]. Because the N_{TP} calculation for packed towers is identical to that for trayed towers, the materials in the trayed sections above are also helpful. The equilibrium stage numbers can be adjusted to match up the measured compositions.

Verified computer programs based on rigorous equilibrium-stage calculations with plate-to-plate material and energy balances should be used to determine N_{TP} whenever possible. When an integral N_{TP} does not result in a close match to the key component compositions in the product streams, a fractional number of stages can be calculated by the method of plotting the computer-predicted percentage recovery of a key component versus the number of theoretical stages described in Section 605.1. Similar to trayed columns, the rate-based model can also be applied to packed bed sections for back-calculations of N_{TP} (and HETP) from the vapor and liquid composition profiles.

605.2.1.3 HETP Assumptions: Because the theories of N_{TP} are well developed, and the tools for N_{TP} calculation are readily available, the HETP model is popular for packed tower efficiency analysis, especially for distillation systems, including multi-component systems and complex petroleum systems. However, three assumptions inherent in this model should be fully understood when using it [61].

The first assumption is that HETP values are constant. This is not necessarily the case. Over the height of a packed bed, the HETP model simply assumes that the vapor and liquid streams leaving a theoretical stage are in equilibrium, giving no quantitative index on how difficult this equilibrium is to establish. However, in reality, as the driving force between the two streams varies, the HETP also varies from stage to stage across the bed. Furthermore, at different process conditions, the HETP values for a packing material may also be different [62, 63].

HETP is typically assumed to be independent of the components that are present, and their relative volatilities. However it could well be the case that the HETP for the key components might be one value, while the HETP for trace components is considerably different. If it should become important to understand where those trace components go with accuracy, say for emissions calculations, this assumption can cause problems.

The second assumption is that mass transfer occurs stage-wise. Fundamentally, the stage-wise concept of HETP does not apply very well to the continuous differential performance of a packed bed. The HETP model assumes that each equivalent theoretical plate works as a perfect mixing zone, even though the two leaving streams are at different elevations of the packed bed. While the vapor and liquid streams at the same elevation in a packed bed influence each other, conditions of vapor and liquid streams at different elevations, (e.g., the vapor and liquid streams leaving a theoretical stage) cannot truly be coupled together by the equilibrium relationship. Also, whenever an external stream different in thermal condition or composition from the internal stream conditions is introduced into a packed bed, it may not be correct to assume that the two streams have reached equilibrium in one HETP may not be correct.

The third assumption is that the gas and liquid streams in each individual HETP have the same temperature that can be used to estimate all the transport and equilibrium properties of that stage. Actually the temperatures of these two streams may differ greatly.

Because of these necessary assumptions, use of the HETP concept should be limited to relative performance, where the main objective is to compare the efficiencies of various types of packing at similar process conditions, but not to obtain a generalized HETP correlation for a specific packing material.

In summary, the HETP concept is easy to understand and use, but the values analyzed for a packing at a given process condition may not be applicable for other process conditions.

605.2.2 *The HTU Model:* The HTU model is derived directly from the two-film theory [64] of mass and heat transfer, along with a material balance around an incremental section of a packed bed [65]. The basic form of the HTU model is:

$$h_T = (HTU) \quad (NTU) \qquad \qquad \textbf{(605.2.2.1)}$$

Because this model treats a packed tower as a continuous differential contactor, there is no need to assume that the gas and liquid temperatures in any horizontal section be the same. Conditions around the feed section where an external stream, whose composition and thermal conditions are significantly different from the internal streams, can be properly simulated.

Conventionally, the HTU model is used only in analyzing the mass transfer of a single component across the gas liquid interfaces. Various stripping and absorption applications can be adequately analyzed using this single-component mass transfer approach. For multicomponent mass transfer applications, the work of Billingsley [66], Krishnamurthy [67,68], Sivasubramanian [69-70], Seader [71], and Taylor [21] should be consulted. Because of the complexity of the HTU model for multiple component simulations, the model validation time required for such a simulation is usually longer than the respective HETP simulation.

The processes commonly analyzed by the HTU model can be classified into three categories:

- Gas phase mass transfer limited processes;
- Liquid phase mass transfer limited processes; and,
- Gas and liquid phase mass transfer limited processes.

Further discussion of these three categories of processes can be found in Ludwig [72], Perry [73], and Leva [74]. In practice, the systems most commonly analyzed by the HTU model are absorbers and strippers. The HETP model is generally used in distillation.

Two additional processes, which frequently use packed columns, can also be analyzed by modified HTU models. They are:

- Liquid-phase chemical reaction dominant processes; and
- Simultaneous heat and mass transfer processes.

Since each of these processes has its preferred form of the HTU model, the nature of the process should be established before an analysis to ensure that the correct form of the model is used.

A sample NTU calculation appears in Section 802.2 and in Sherwood, Pigford, and Wilke [51].

605.2.3 *The Relationship Between HETP and HTU:* For the special case of binary systems with straight equilibrium and operating lines and with negligible heat effect, a theoretical relationship exists between HETP and HTU:

$$HETP = \left(\frac{ln\{\lambda\}}{(\lambda)-1} \right) HTU_{OG} \qquad (605.2.3.1)$$

$$HETP = (\lambda) \left(\frac{\ln\{\lambda\}}{(\lambda)-1} \right) HTU_{OL} \qquad (605.2.3.2)$$

where

$$\lambda = \frac{m\,G_M}{L_M}$$

(605.2.3.3)

For cases where the equilibrium and the operating lines are straight and parallel, the relationship is:

$$HETP = HTU_{OG} = HTU_{OL}$$

(605.2.3.4)

Equations 605.2.3.1 and 605.2.3.2 imply that the HETP values are a strong function of these slopes. Distillation of non-ideal mixtures, especially at total reflux, is a typical process with significant variation of these slopes. This contributes to the variation in the HETP values reported for distillation processes [62].

605.2.4 *Mass Transfer with Simultaneous Heat Transfer:* Simultaneous heat and mass transfer processes cannot be properly handled by the HETP model, because this model assumes that the temperature of the gas and liquid streams reach equilibrium in one theoretical plate. A modified HTU model or a modified heat exchanger model better simulates these processes. Alternatively, this process can potentially be handled by commercially available non-equilibrium or rate-based simulation packages.

605.2.4.1 The Modified HTU Model: This model uses the gas phase enthalpy, a factor that combines the mass and the thermal conditions of the streams, instead of using gas phase composition as the transfer driving force. The heat and mass transfer coefficients are related by the Lewis number. Theoretically, this model is limited to the simulation of one condensable component. The basic theory of this model was developed by McAdams [75] and Mickley [76]. A good calculation sample can be found in Coulson and Richardson [77], and some engineering data for commercial packings analyzed with this model can be found in Eckert [78] and McNulty and Hsieh [79].

605.2.4.2 The Modified Heat Exchanger Model: In this model, a packed tower is treated as a direct contact heat exchanger. The efficiency of the packing is analyzed by the overall heat transfer coefficient, with some modifications to account for the effect of mass transfer (i.e., evaporation or condensation).

The working equations can be found in Fair [80, 81]. As shown, the well-known Ackerman coefficient is used to correct the effect of mass transfer to the heat transfer coefficient.

The application of this heat transfer model should be limited to those processes with relatively straight gas and liquid temperature profiles, and these two profiles should not cross each other. For other cases [82], the HTU model should be used.

605.2.5 *Packing Efficiency Prediction:* Despite the wide and prolonged use of packing, estimating mass transfer performance efficiency is still heavily dependent on experience or rules of thumb, which set reasonable expectations for many systems of industrial interests. However, those rules of thumb are often not applicable for efficiency predictions to situations that have not been experienced before, because the detailed hydraulics in a packed bed are not completely understood. Furthermore, studies have shown that initial liquid distribution is more critical to packing performance than had been realized, casting some doubt on past data upon which correlations were based. Methods for prediction of random packing efficiency are given by Perry [1] and Billet and Schultes [83]. An interpretation of current methods for prediction of performance of structured packings is given by Fair, et al. [84].

At present, efficiency prediction is made most reliably by comparison with large-scale data from the manufacturers or from a testing service. Comparisons should be made for the packing in question using very similar systems and operating conditions.

700.0 INTERPRETATION OF RESULTS

This section covers sources of error, and gives a checklist of possible reasons for the tower not performing as designed, as well as a procedure for converting test results to projected design conditions. Typical questions of test data interpretation include the following:

- o Was the distillation system able to make the required splits, purities, recovery and capacity? If not, why?

- o What might be done to correct this problem?

- o If the design feed composition and rate were not used during the test, what splits should the tower make with the design feed composition and rate?

701.0 Sources of Experimental Error

Commercial scale plant tests will normally involve greater errors than those undertaken in research laboratories, where equipment and instrumentation are designed specifically for test conditions [7, 20, 32, 85-88].

The following are potential causes of experimental error in commercial plant testing:

a. sampling
b. sample analysis
c. interpretation
d. the presence of unsteady state

These potential errors should be minimized to the extent practical for best test results.

One cause of incorrect sampling can be misinterpretation of column performance. Sampling problems can be:

a. too few sample points available
b. non-calibrated standard samples
c. unreliable sample technique
d. unpurged sample lines
e. inadequate location of sample points
f. incorrect sample labels
g. leaking sample lines
h. sample contaminations

Errors can also occur due to instrumentation, for example, wrong flow rate, temperature or pressure drop reading.

Test data interpretation errors may be introduced by use of incorrect data on the following items: (a) enthalpy, (b) latent heat, (c) vapor pressure, (d) "K" values, relative volatilities or VLE data, (e) assumption of no heat of mixing, (f) presence of water or other impurities, (g)

density of liquid and vapor, (h) viscosity of liquid and vapor, (i) surface tension and (j) tray or packing efficiency (leading to inaccurate simulation of the separation).

If unsteady state or cycling occurs during the test runs, the results may be meaningless, or at best very difficult to interpret. Lack of steady state may be caused by: (a) insufficient time to line out column, (b) varying feed rate and/or compositions (e.g., feed coming directly from a reactor or another column), (c) varying feed enthalpy or percent vaporization, (d) tower cycling due to instrumentation system, poor instrument tuning, or improperly sized valves, (e) improper control points (temperature, pressure), (f) in hydrocarbon distillations, accumulation of water in column or inadvertent refluxing of water into the column, (g) disturbances in heating or cooling media transmitted to the column (e.g., changes in steam pressure, cooling water temperature, etc.), (h) chemical reaction and (i) disturbances in heat losses, i.e., rain storms.

701.1 Material and Enthalpy Balances

As discussed in Sections 602.0 and 603.0, adequate closure of material and enthalpy balances is important for proper interpretation of distillation test data.

701.1.1 *Inability to Close Material Balances*

a. Overall Balance-Typical causes for poor closure of overall material balances are: (a) flow meter errors, (b) density errors, (c) hold-up changes in the tower, exchangers or reflux drum, (d) vent losses or leaks, (e) heat exchanger leaks into system, and (f) unsteady operations.

b. Component Balance-In addition to those causes listed above are the following: (a) non-representative samples taken or poor sample technique, (b) incorrect analysis of samples (c) insufficient number of samples taken to be statistically significant, (d) column not at steady state or inventory changing with time (e) heat exchanger leaks, (f) presence of unknown components, (g) misinterpretations of compositional units.

Incorrect analysis of samples can have several aspects. One is flaws in the analysis technique. Another is misidentification of compounds, particularly for distillations that give a very pure product. Misidentifying an unknown or trace compound as a small amount of one of the main compounds is common and can interfere with the understanding of efficiency. Another is improper calibration between weight percent and area percent on these gas chromatography scans.

701.1.2 *Inability to Close Enthalpy Balances*

Typical causes for poor closure of enthalpy balances are: (a) poor material balance closure, (b) errors in temperature, pressure, composition or enthalpy of feed or product streams, (c) errors in reflux flow rate, temperature, or composition, (d) entrained liquid from the top of the packed bed or tray to the reflux accumulator, (e) errors in flow rate, temperature, pressure, or composition of heating or cooling fluids, (f) defective steam traps, (g) inaccurate enthalpy data or correlations for process, heating or cooling streams, (h) improper assumptions regarding radiation or convection heat losses or gains.

702.0 Effects of Experimental Error

Using computer simulation tools to model the column separation and to identify experimental errors is recommended. This can be done by introducing small changes to the feed and reflux rates, reboiler duties or composition data from the test run and determining the effect on theoretical efficiency or on calculated internal load. The effect of erroneous vapor/liquid equilibrium data can be analyzed in the same way [89].

703.0 Design versus Performance

Models for predicting pressure drop and separation efficiency, plus guidelines for selecting packings or trays with optimum pressure drop per theoretical stage are available in the literature [46,47,83,90-96]. Beside a literature survey, most trays and packing vendors have programs available to estimate expected pressure drops of their products.

Hydraulic performance of operating columns is determined by examining the throughput and pressure drop measurements. The mass transfer efficiency of a packed or trayed column is adversely affected by poor hydraulic performance so the first consideration is to ensure the column is operating hydraulically correct within the allowable loading range. The mass transfer separation, as required in a plant operation, is commonly specified as a desired yield or purity of top, sidestream or bottom product.

If the column operation yields unexpected separation and/or poor hydraulic performance, the major areas to be checked are as follows.

703.1 Mechanical/Tower Equipment

In case design capacities cannot be met, the following causes should be investigated.

    a. inadequate condensate removal due to inert gases or undersized condenser, reboiler or heat exchanger
    b. pump problems
    c. insufficient quality or amount of utilities
    d. instrumentation or control problems
    e. insufficient feedstock quality or amount
    f. piping design or installation errors
    g. mechanical blockages (plugging of trays or packings)
    h. equipment damage
    i. errors in equipment location
    j. equipment fouling
    k. improper modifications or retrofitting
    l. distributor overloading or underloading
    m. spacing of tower internals

For more detailed information, refer to the literature [85, 87, 97-98].

703.2 Process Conditions

The major process indications of not meeting design conditions are:

    a. tower flooding before reaching design capacity.
    b. actual pressure drop significantly higher or lower than calculated.

    c. inability to make design splits.

    d. cycling of temperature and inability to reach steady state [97].

    e. column temperature gradient different from computer simulation.

    f. condenser or reboiler unable to meet design duties .

    g. existence of pinch points.

    h. improper distribution of liquid and vapor (flashing, foaming, plugging).

    i. inability to operate tower at design pressure.

    j. inaccuracy of equilibrium calculations.

These problems are further detailed in Sections 704.0 and 705.0. Additional troubleshooting procedures in operating columns are given in [7] and [99-110].

704.0 Hydraulic Performance

A column malfunction is most often indicated by an unusual hydraulic performance, with premature flooding or excessive pressure drop. The most common reasons for this are listed below.

704.1 Mechanical/Tower Equipment

    a. Poor fabrication

    b. Dirt, fouling, corrosion products or other deposits

    c. Freezing components

    d. Poor design or installation of tower internals, e.g., reflux or feed lines, reboiler return, internal baffles, chimney trays, side stream draw boxes or nozzles and sumps

    e. Leaking exchangers or water entering with feeds (for hydrocarbon systems)

    f. Feed composition or enthalpy out of design

    g. Poor design of liquid outlet lines (i.e., vortexing or lack of self-venting)

    h. Too little free cross sectional area (i.e., for vapor/liquid distributor)

    i. An unexpected bottleneck in the column due to:
        1) damaged internals.
        2) superficial area decreased,
            as in a section of old tray rings.

    j. Internals installed upside down or improperly or a liquid distributor too close to the packing.

    k. A distributor overloaded.

    l. Poor design of hold-down and/or support plate.

704.2 Tray

    a. Poor tray design or installation

    b. A tray or downcomer blockage or fouling, such as lunch pails in downcomer

    c. Downcomer exit and/or entrance area too small

    d. Incorrect valve sizing or insufficient open area

    e. Omission of anti-jump baffles on trays with high liquid loading and narrow center downcomers

    f. Poor design or installation of internals, e.g., calming zone, weirs, downcomers, transition trays, seal pans, multipass trays, and downcomer trapouts

    g. Maldistribution on multipass trays

    h. Excessive entrainment

    i. Vaporization in downcomers

    j. Insufficient vapor disengagement in downcomers

    k. Unsealed or improperly sealed downcomers

    l. Excessive downcomer backup

    m. Excessive liquid loads

    n. Damaged trays or internals

For more detailed information, refer to the literature [7, 87, 97, 108, 111-123]

704.3 Packing

    a. Poor installation

    b. Inaccurate characterization of packing factor

    c. Breakage, compression or settling

    d. Corrosion, rust or thermal degradation

    e. Deposits of particles or foulants or biological growth

    f. Poor packing selection (wrong size, material)

704.4 Process Conditions

    a. Inadequate boil-up rate or condenser/cooling capacity

    b. Control system cycling

    c. Obstructions in piping or underrated pump capacity

    d. Cycling of steam or cooling rates

    e. Foaming

    f. Impurities leading to two-phase liquid mixtures, e.g., free water phase in light hydrocarbon towers

    g. Plugging of internals due to solids buildup

The location of the flooding, blockage or choke point is usually determined by use of pressure drop measurements, column temperature profiles, or radiation scanning. Cause of localized flooding is most commonly diagnosed by reviewing tower internals designs and external piping circuits. The possibility of leaking exchangers may be diagnosed by appropriate pressure, vacuum, or leak tests using radioactive tracers. Entrainment may be checked by injection of dyes or tracers into the feed, reflux, or reboiler return. Gamma scanning is a proven technique to troubleshoot column malfunction (see Section 309.2).

The possibility of two liquid phases (water) is normally checked by calculations on the suspected column section, using the best VLE and LLE data available, together with computer simulations at test run conditions.

Foaming can be checked in the field using gamma scan, level gages on the tower, or by small-scale tests using real tower liquids over the operating temperature and pressure range, and bubbling gas through the liquid in the plant laboratory, or suddenly depressuring the boiling liquid. If foaming is noted in the small test, it may normally be solved by continuous injections of an effective defoaming compound.

Maldistribution is especially possible on multi-pass tray designs or in packed columns. Maldistribution is most easily checked by tower temperature gradients taken circumferentially around the tower or by radiation scanning (see Section 309.2).

705.0 Mass Transfer Performance

Mass transfer efficiencies in columns are evaluated from the measured compositions and are expressed in terms of overall efficiency for trayed towers, and HETP or HTU_{OG} for packed towers. The results are compared with the original design efficiency obtained from proprietary data, correlations, or from experience, to decide if the column is operating within specifications. A partial list of available mass transfer models and published experimental data are given in [46, 83, 95, 96, 138]. Predicted mass transfer efficiencies have a typical error of $\pm$ 20-30% for systems with thermodynamic and physical properties close to those of the experimental test systems. The error can be higher if the separation system has properties far different from typical test systems.

If commercially measured mass transfer efficiencies appear lower than expected, some possible causes are listed below.

705.1 Mechanical/Tower Equipment

a. Poor internal design for side draws, reflux, feed, or reboiler piping
b. Incorrect thermocouple/thermowell installation or calibration
c. Wrong type of liquid distributor
d. Wrong type of gas distributor
e. Wrong flashing feed device
f. Wrong type of, or lack of, mist eliminator
g. Problems with upstream or downstream equipment
h. Unsuitable design of condenser, reboiler, and other auxiliaries and control valves
i. Incorrect calibration of instruments
j. Blockages or malfunctions in control valves or instrument taps or orifices, pressure drop instruments, thermocouples, or level controls
k. Incorrect piping
l. Exchanger leaks
m. Leakage of spare feed point and block valve

For more detailed information, refer to the literature [88, 97, 100, 123, 139-140].

705.2 Tray

a. Incorrect assumption of tray efficiency
b. Poor tray and internals design, fabrication, or installation
c. Damaged, corroded, or fouled trays or internals
d. Unsealed downcomers or inadequate seal pans
e. Maldistribution of vapor or liquid
f. Nonlevel trays, weirs, or downcomer seals
g. Excessive tray leakage or weeping
h. Underloading on trays
i. Excessive entrainment
j. Vapor bypassing of tray liquid

    k.   Excessive liquid recirculation on trays

    l.   Excessive shearing of tray liquid to form "fog" or spray on trays, which occurs at low liquid rates, combined with a high gas rate. As a result, liquid is not able to flow down the downcomers, a process also called "blowing the trays dry."

    m. Use of tray types that are too rate sensitive, particularly where tray traffic varies widely from top to bottom of the column, such as in demethanizers, depropanizers, extractive distillation, or vacuum columns

    n.   Two liquid phases on trays

705.3 Packing

    a.   Low-pressure drop and inadequate effective surface area for vapor-liquid contact. This may occur if packing is too large compared with the small column diameter.

    b.   Excessive pressure drop per theoretical stage, due to flooding, fouling, foaming, or damage

    c.   Poor wetting

    d.   Wrong material

705.4 Maldistribution

Maldistribution is non-uniform distribution of gas or liquid over a column cross-sectional area.

A large-scale maldistribution can be caused by:

    a.   extensive plugging of distribution points in one or more large areas of a distributor

    b.   extensive plugging of one or more large packing zones/packing volume

    c.   extensive plugging of one or more large tray or downcomer zones

    d.   tilted liquid distributor

    e.   tilted tray

    f.   tilted column

A small-scale maldistribution can develop from:

    a.   random blockage of a few distribution points in a distributor

    b.   local plugging of packing, tray or downcomer zones

If no large-scale liquid maldistribution exists, a gas maldistribution is often self-corrected at a certain height of the packed bed, or after a few trays due to the continuous vapor phase flow and the vapor pressure drop. In contrast, a liquid maldistribution is more difficult to be corrected by the tray or packing itself and often has a more severe impact on the separation efficiency.

Often the failure to produce the desired separation in multi-pass trays or a packed column can be traced to gas or liquid maldistribution. Maldistribution should be suspected if the same tray or packing (type and size) with the same system has produced an acceptable overall separation in a smaller diameter column, or has developed a higher efficiency or smaller overall HETP (or HTU) in a shorter bed.

Controlled maldistribution studies on packed columns show that large-scale maldistribution produced more detrimental effects on overall separation than small-scale maldistribution. [133,141-143]. To interpret this, Zuiderweg has postulated a "Zone-Stage" model for packed columns in which a liquid distribution profile through the bed results in bulk flow between zones with resulting variation in efficiency. FRI has confirmed that this does, in fact, explain some of their results [142-144].

Another approach to study the effect of maldistribution on trayed and packed column efficiency is to split a column into several columns in parallel. The parallel column sequences are fed with liquid and gas rates with a plus/minus variation factor, apart from the mean flow rates. Process simulators can be an effective way to apply this technology [145-146].

Usually, the models predict that the effects of maldistribution are worse for operations close to pinch points, where operating line meets equilibrium line; low relative volatilities; very long beds; or when attempting to develop a large number of theoretical stages in one bed. Similar results have been reported by Moore and Rukovena [147] and by Perry [148].

If maldistribution is the cause of poor performance of a trayed or packed column, the four possible sources are:

a. inadequate design of the (multipass) tray deck (flow path length and/or pressure drop) or inadequate design of liquid distributor for packed columns [149-153].
b. improper installation of trays or packing or the distributor.
c. problems occurring during operation (fouling).
d. vapor maldistribution.

For these reasons, the analysis of a suspected distribution problem should start with a review of the operating history to decide if the problem was always present or if it developed during operation. If the problem developed during operation, it should be determined whether it developed gradually during a period of reasonably steady process operation, if its appearance coincided with changes in process conditions, or if it developed suddenly without process changes. In the latter case, the prime suspect is some sort of mechanical failure and little recourse exists other than to shut down, open up and inspect the column and internals.

Fouling should be suspected if the problem developed slowly during steady operation, particularly if small openings (5/32″ [4 mm] or less) are on a tray deck or in a distributor. Rust particles have been known to plug small orifices between start-up and the achievement of steady state.

If the appearance of the problem coincides with the changes in process operation, the hydraulics should be analyzed for possible causes, such as multipass tray balancing, overflow of distributors, and/or excessive feed flashing. Note that the extent of feed flashing must be looked at on a volumetric, and not a weight basis. Suitable tests can then be devised with the unit on stream, for example, by reducing the feed temperature to lower the amount of vapor in the flashing feed.

Specific causes of maldistribution can include:

a. poor installation of packings or tray.
b. balancing problems with multipass trays.
c. non-uniform flow, which may lead to localized intersection of the operating and equilibrium lines causing premature pinch conditions.
d. unexpected presence of vapor in the feed.
e. development of excessive wall flow in packed columns due to improper liquid distribution.
f. too few distribution or pour points.

g. distributor capacity that does not ensure uniform flow, due to liquid head being too low, or distributor overflow.
h. distributor plugged, corroded, or designed, fabricated, or installed improperly.
i. poor vapor distribution.
j. excessive aeration of distributor liquid.
k. inadequate redistribution.

Maldistribution can often be detected by differences in temperature around the periphery of the column.

705.5 Process

Process design related factors can also cause a reduction in column efficiency, including:

a. incorrect VLE data used in design.
b. non-representative samples.
c. too few samples to be statistically significant.
d. incorrect sample analysis.
e. incorrect enthalpy or physical properties used in design.
f. non-design feedstock, composition or enthalpy.
g. steady state not reached or tower cycling.
h. non-optimum feed tray/stage.
i. poor column control.
j. incorrect tower operating conditions.
k. incorrect utilities or heating and cooling condition.
l. incorrect reflux temperatures.
m. incorrect overhead draw rate.
n. incorrect instrument installation, leading to incorrect readings or control.
o. excessive heat leaks.
p. air leaks in vacuum towers.
q. contamination of the column feed (e.g., water).
r. azeotropes.
s. foaming.
t. incorrect calculation procedure for number of trays/stages or invalid critical assumptions.
u. imbalance of rectifying and stripping stages, or unusual behavior, upon analysis.
v. excessively cold or superheated feed.
w. chemical reaction in the column.
x. incomplete mixing of the liquids entering the feed section.
y. column not at steady state.

For more detailed information, refer to the literature [7, 18, 85-87, 97-98, 114-118, 123, 137, 139-140, 154-162].

A procedure for diagnosing the reasons for column malperformance must start with the questions: "Is the problem real?" and "What is the magnitude?" A logic diagram for distillation column troubleshooting (Figure 3) can be extremely useful. Finding the problems and developing economical short-term and longer-term rapid solutions is usually better than spending weeks achieving extreme accuracy in defining the tray efficiency found in the test. Comparison of the deficient tower with successful operating towers is also frequently used.

706.0 Test Troubleshooting

Apparent poor performance may actually be caused by problems with the test procedure. Some common problems are:

706.1 Analysis Procedure

a. Accurate measurement of compositions is required for correct efficiency computation. Large errors are possible if only the bottoms and the overhead product are used to determine efficiency.
b. Analyses must be frequently checked against calibrated sample and blank sample to insure proper analytical accuracies.
c. Samples may contain carryover contamination.
d. Detector may have low sensitivity at low or high concentrations.
e. Sample handling technique must be reliable and repeatable

706.2 Sampling

To reduce possibility of sampling errors, follow the guidelines below.

a. Purge sample lines until hot process fluid flows.
b. Exclude vapor from liquid samples.
c. Ensure sample lines are not leaking.
d. Ensure sample locations are representative.
e. Ensure vapor sample lines have no low points for collecting liquids.
f. Confirm sample lines or labels are not interchanged.

706.3 Equilibrium Data

Equilibrium data errors can be caused when:

a. the ideal equilibrium model (Raoult's Law) is often not of sufficient accuracy.
b. proper equilibrium model to predict non-ideal liquid or vapor phase behavior is not selected.
c. the system may be partially miscible, such as water present in organic systems.
d. inert components, unexpected contaminants, or hydrogen may be present.
e. higher precision is required, as when the relative volatility or K-value approaches unity.

706.4 Temperature Measurements

Temperature measurement errors can be caused when:

a. sensors are not calibrated properly with ice slurry and/or boiling water.
b. thermowells may be cracked, thermocouples may be touching metal, or connection of leads may be reversed.
c. shielding may be incorrect or noise filtering inadequate.
d. non-uniform temperature profiles may indicate non-uniform flow and separation.

706.5 Heat and Material Balances

Heat and material balance data errors can be caused when:

a. flow measurement device is not calibrated correctly or exceeds linear range.
b. heat losses are excessive.
c. composition or temperature measurements are incorrect.
d. heat exchanger and process stream inventory changes.
e. measured reflux or vapor rates do not match. Check with rates obtained through duty checks.

706.6 Fluctuation of Process Conditions

Fluctuations in process condition can also influence the accuracy of test results. Process fluctuations can occur when:

a. steady state operation is not yet achieved.
b. faulty control causing cycling of unsteady state operation.
c. process feed varies from specifications.
d. cooling water and steam flow rates fluctuate.

706.7 Pressure Drop Measurements

Well-located pressure drop measurements or temperature profiles [31] are useful in determining where poor or abnormal performance occurs in the packings or trays due to:

a. plugged tap lines or liquid in tap lines.
b. broken or disconnected line connections.
c. connections that are too small (1/2 inch or 13 mm minimum).
d. limited sensitivity (should be calibrated to <0.1 inches of water/ft [<1 mbar/m] range, less for low pressure drop packings).
e. errors due to steep/fluctuating responses at high vapor rates (i.e. flooding).

706.8 Incorrect Prediction of Pressure Drop

Errors can also occur when incorrect predictions of pressure drop are made, such as when:

a. packing is not accurately characterized due to variations in vendor specifications or manufacturing.
b. viscous fluids effect are present.
c. there are variations in physical properties throughout the column.
d. there is foaming.
e. there is a poor predictive model or incorrect extrapolation.
f. static gas head for high-pressure systems are omitted.

706.9 Errors in Assumptions in Modeling Mass Transfer

Common assumptions in mass transfer modeling can also lead to errors, such as:

a. incorrect or varying properties through column.
b. calculations extended beyond reliable limits.

    c.   scale-up of small column data is not reliable.

    d.   model is not based on data for similar chemical systems.

706.10 Multicomponent Systems Deviate from Binary Data

Multicomponent techniques for evaluating mass transfer performance are given in [21, 109, 135-136, 163].

    a.   Mass transfer interaction between components is not properly modeled for multicomponent system.

    b.   Multicomponent equilibria techniques are inaccurate or unavailable.

    c.   Significant heat transfer between phases impacts mass transfer (rate based modeling recommended).

706.11 High Purity Separation

Simulation results of high purity separations are very sensitive to the selection of the equilibrium model. Validating an equilibrium model against experimental data or simulating the process with various available vapor/liquid equilibrium models to check the sensitivity of the results (number of theoretical stages required for high purities) is recommended. Important components should be selected to estimate the error between simulated and measured compositions. This error in the component balance is recommended to be less than 10% for these important components.

706.12 Test and Design Conditions

Frequently the conditions for the test cannot be the same in all respects as those for which the equipment was designed. If conditions reasonably close to design can be achieved, a test can still be made to check the column, and test results can be extrapolated to the design conditions by techniques based on sound engineering principles. Usually, the procedure would be as follows:

    a.   Using the design VLE and enthalpy data and taking into account the experimental uncertainties, calculate the maximum and minimum numbers of theoretical stages in the rectifying and stripping section under test conditions.

    b.   Assume that the same number of theoretical stages will be developed under design conditions (i.e., that the tray efficiencies will be the same at design conditions as they were for the test conditions).

    c.   Using these numbers of theoretical stages, calculate by computer simulation the best and worst splits of the key components under design conditions.

This technique may be particularly applicable for interpreting results for an acceptance test (Section 305.2).

800.0 APPENDIX

801.0 Notation

A_A	Active area of tray, (ft^2, m^2)
A_{dow}	Cross sectional of downcomer (ft^2, m^2)
AF	Aeration factor used in estimation of sieve tray pressure drop
A_h	Hole area, (ft^2, m^2)
C_L	Liquid capacity factor (gpm/ft^2, m^3/m^2-hr) (Figure 15)
CP	Heat capacity (Btu/lb-°F, kJ/kg-°C)
C_S	C-factor $\left[= U_S \sqrt{\dfrac{\rho_V}{\rho_L - \rho_V}} \right]$ based on tower superficial cross-sectional area (ft/s, m/s)
C_{sbf}	Corrected flood C-factor (ft/s, m/s) (Figure 20)
D	Distillate Flowrate, (lb/hr, Kg/hr)
F_h	Vapor hole kinetic head, ($m/s\sqrt{kg/m^3}$)
F_P	Packing factor (ft^{-1}, m^{-1})
FP	Flow parameter
FPL	Flow path length, (in, m)
G	Gas mass velocity (lb/hr-ft^2, kg/hr-m^2)
G_M	Molar gas mass velocity (lb mole/hr-ft^2, g-mole/s-m^2)
HETP	Height equivalent to a theoretical plate (ft, m)
h_D	Clearance under downcomer (in., mm)
h_D'	Hydraulics gradient on tray, (in., mm)
h_{dry}	dry head loss used in tray pressure drop calculation (in of liquid, mm of liquid)
h_{DS}	Calculated clear liquid height (in., mm)
h'_L	Head loss associated with the wet pressure drop on tray (in of liquid, mm of liquid)
h_{ow}	Height of crest over weir (in., mm)
h_T	Total depth of packed bed (ft, m)
h_{total}	Total tray pressure drop (in of liquid, mm of liquid)
H_{VAP}	Vapor enthalpy (Btu/lb, kJ/kg)

h_w	Weir height (in, mm)
HTU	Overall height of a vapor (HTU_{OV}), liquid (HTU_{OL}), or gas (HTU_{OG}) phase transfer unit (ft, m)
K	Constant in dry tray pressure calculation
L	Liquid mass velocity (lb/hr-ft^2, kg/hr-m^2)
L_M	Molar liquid mass velocity (lb-mole/hr-ft^2, g-mole/s-m^2)
L_w	Weir length used in pressure drop calculation (in, m)
m	Equilibrium line slope
N_{TP}	Number of theoretical plates
NTU	Number of overall vapor (NTU_{OV}), liquid (NTU_{OL}), or gas (NTU_{OG}) phase transfer units
P	Total pressure (psia, torr)
ΔP	Total pressure drop, (in. of H_2O/ft or in. of H_2O, mmHg/m or mmHg)
ΔP_{dry}	Dry tray pressure drop (in. of liquid, mm Hg)
ΔP_{total}	Total pressure drop (in. of liquid, mm Hg)
Q	Heat transferred (Btu/h, kW)
Q_{Bot}	Bottoms enthalpy (BTU/hr, kJ/hr)
Q_{Cond}	Heat removed by the condenser (BTU/hr, kJ/hr)
Q_{Dist}	Distillate enthalpy (BTU/hr, kJ/hr)
Q_{Feed}	Feed enthalpy (BTU/hr, kJ/hr)
Q_{in}	Energy entering distillation control volume (BTU/hr, kJ/hr)
Q_L	Volumetric liquid flow (gpm or ft^3/h, m^3/h or m^3/s)
Q_{Loss}	Heat loss (BTU/hr, kJ/hr)
Q_{out}	Energy leaving distillation control volume (BTU/hr, kJ/hr)
$Q_{Reboiler}$	Heat supplied by the reboiler (BTU/hr, kJ/hr)
Q_{Reflux}	Enthalpy in reflux (BTU/hr, kJ/hr)
Q_V	Volumetric vapor flowrate (ft^3/hr, m^3/hr)
Q_{vapor}	Vapor enthalpy entering condenser (BTU/hr, kJ/hr)
R	Reflux ratio
T	Temperature (°F, °C)
T_{Dist}	Temperature of the liquid leaving condenser (°F, °C)
T_{Vap}	Temperature of vapor entering condenser (°F, °C)
U_{nf}	Net vapor velocity at flood used in Fair flooding chart (m/s) (Figure 20)
U_S	Superficial Vapor velocity, (ft/s, m/s)

V	Vapor velocity (ft/s, m/s)
V_{dow}	Downcomer velocity (ft/s, m/s)
V_h	Vapor hole velocity (ft/s, m/s)
W_G	Vapor mass flow rate (lb/h, kg/h)
W_L	Liquid mass flow rate (lb/h, kg/h)
X	Composition in liquid (mol fraction)
X_{Bot}	Liquid composition (mole fraction) in bottom product
X_{Dist}	Liquid composition (mole fraction) in distillate
X_{Feed}	Liquid composition (mole fraction) in feed

801.1 Greek Symbols

α	Relative volatility
ν	Kinematic viscosity of liquid $(= \mu_L/\rho_L)$, (cS)
λ	Stripping factor (slope of equilibrium line)/(slope of operating line)
ρ_L	Liquid density (lb/ft^3, kg/m^3)
ρ_V	Vapor density (lb/ft^3, kg/m^3)
μ_L	Liquid viscosity (cP)
σ	Surface tension (dyne/cm)

802.0 Sample Calculations

802.1 General Analysis of Test Data

Section 802.0 provides sample calculations for the analysis of plant test data. These steps include determining the quality of the data and evaluating the performance of the column internals (packing or trays). The calculations presented in this section should be performed prior to evaluating data. Sample calculations involve setting up a process flow diagram of the distillation column, followed by stage-to-stage process simulation for additional analysis. The calculations are presented with both engineering and SI units. These conversions can be found in *Perry's Chemical Engineer's Handbook* [1].

Additional internal performance calculation methods may be found in these reference books:

Perry's Chemical Engineer's Handbook, Chapters 13 and 14 [1]

Kister, H.Z., *Distillation Operations* [6]

Kister, H.Z., *Distillation Design* [46]

Other references are provided in the packed column and trayed column sections (802.2 and 802.3).

802.1.1 *Material and Enthalpy Balances:* Obtaining reasonable closure of material and energy balances is important. The quality of all test data should be verified by determining the overall and component material balances. To do so, it is necessary to acquire steady state flow rate and component compositional data, and to perform an energy balance to determine internal liquid and vapor flow rates, or to verify a computer simulation. Preparation of a process flow diagram of the distillation system is also recommended.

802.1.2 *Column Performance Determination:* The performance of each column section should be checked to verify proper operation. After obtaining successful material and energy balances, the internal liquid and vapor rates can be determined for the rectifying and stripping sections using a process simulator.

802.2 Packed Column

802.2.1 *Packed Column Example Calculations with English Units:* A process simulation was performed using the measured reflux ratio, flow rates and pressures. The results are shown in Table 5. A column of 15 theoretical plates, with the feed on the 11th plate, most closely matched the data.

Column Dia (ID)	48"	
Packed Sections	Rectifying	Stripping
Packing Type	Slotted Rings	Corrugated Sheet Metal
Packing Size	2"	No. 2 (200 m^2/m^3)
No. Beds	2 (top and middle beds)	1 (bottom bed)
Height each bed	15 ft	7.5 ft

Table 5. Packed Column Arrangement-O/P-Xylene Distillation (Eng Units)

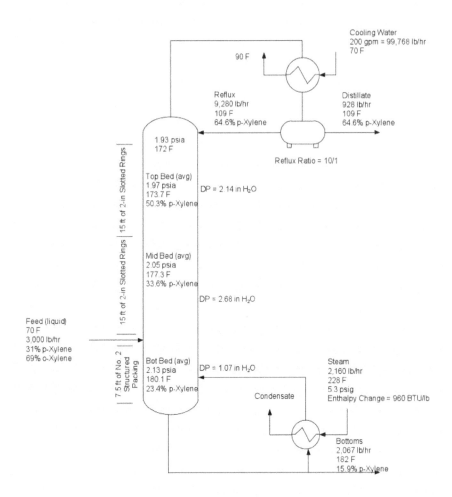

Figure 17. Packed Column Distillation Process Flow Diagram for O/P-Xylene Distillation (Eng Units)

Material Balance

Overall Flow Material Balance

% Error = 100 * (flow in - flow out) / flow in

= 100 * (3000 - 928 - 2067) / 3000 = 0.2% (excellent)

Component Material Balance Check

P-XYLENE

Flow in= 3000 * 0.31 = 930 lbs/hr

Flow out, distillate = 928 * 0.646 = 599.5 lb/hr

Flow out, bottoms = 2067 * 0.159 = 328.7 lb/hr

% Error = (930 – 599.5 – 328.7) / 930 = 0.2% (excellent)

This is within the desired 3%.

Theoretical Stage Number	Temp, °F	Pressure, psia	Flows, lb/hr		Density, lb/ft³		P-Xylene Concentration Liquid Wt. Fraction
			Vapor	Liquid	Vapor	Liquid	
Condenser	109.6	1.93	928*	9,282	--	53.2	0.646
1	172.0	1.93	10,211	10,790	0.0305	52.1	0.596
2	172.9	1.95	11,717	10,786	0.0307	52.1	0.548
3	173.7	1.97	11,714	10,782	0.0310	52.2	0.503
4	174.5	1.98	11,710	10,779	0.0312	52.2	0.461
5	175.3	2.00	11,707	10,776	0.0314	52.2	0.424
6	176.0	2.02	11,705	10,775	0.0316	52.3	0.391
7	176.6	2.03	11,702	10,773	0.0319	52.3	0.361
8	177.3	2.05	11,701	10,773	0.0321	52.3	0.336
9	177.9	2.07	11,700	10,773	0.0323	52.3	0.314
10	178.4	2.08	11,700	10,773	0.0326	52.3	0.295
11	179.0	2.10	11,700	14,618	0.0328	52.3	0.279
12	179.5	2.12	12,550	14,619	0.0330	52.4	0.257
13	180.1	2.13	12,553	14,621	0.0333	52.4	0.234
14	180.8	2.15	12,556	14,624	0.0335	52.4	0.210
15	181.4	2.17	12,557	14,626	0.0337	52.4	0.185
Reboiler	182.0	2.18	12,559	2,067	0.0339	52.4	0.159

*Distillate Flow

Table 6. Process Simulation of O/P-Xylene Distillation (Eng Units)

Enthalpy Balance - Datum Condition: Liquid at 109 ° F

OVERALL

$$Q_{Reboiler} = 2160 * 960 = 2.073 * 10^6 \text{ Btu/hr}$$

Heat Capacity of the Feed = 0.406 Btu/lb-F

$$Q_{Feed} = 3000 * 0.406 * (70 - 109) = -47,502 \text{ Btu/hr}$$

$$\text{Heat in} = Q_{Reboiler} + Q_{Feed}$$
$$= 2.073 * 10^6 - 47,502$$
$$= 2.025 * 10^6 \text{ Btu/hr}$$

Heat Capacity of Cooling Water = 1 Btu/lb-F

$$Q_{Cond} = 99,768 * 1 * (90-70) = 1.995 * 10^6 \text{ Btu/hr}$$

Heat Capacity of Distillate = 0.421 Btu/lb-F

$$Q_{Dist} = 928 * 0.421 * (109-109) = 0 \text{ Btu/hr}$$

Heat Capacity of Bottoms = 0.462 Btu/lb-F

$$Q_{Bot} = 2067 * 0.462 * (182-109) = 69,712 \text{ Btu/hr}$$

Assume Heat Loss is negligible

$$\text{Heat out} = Q_{Cond} + Q_{Dist} + Q_{Bot} + Q_{Loss}$$
$$= 1.995 * 10^6 + 0 + 69,712 + 0$$
$$= 2.064 * 10^6 \text{ Btu/hr}$$

$$\% \text{ Error} = (Q_{in} - Q_{out}) / Q_{in}$$
$$= (2.025 - 2.064) / 2.025 = -1.8\% \text{ (Very Good)}$$

OVERHEAD LOOP

Vapor Rate = (R+1) * D = 11 * 928 = 10,208 lb/hr

Heat of Vaporization = 165 Btu/lb

$$\text{Heat in} = Q_{vapor}$$
$$= 10,208 * (165 \text{ Btu/lb}) + 10,208 * 0.421 * (172 - 109)$$
$$= 1.955 * 10^6 \text{ Btu/hr}$$

$$\text{Heat out} = Q_{Cond} + Q_{Reflux} + Q_{Dist} + Q_{Loss}$$

At Reference Temperature of 109 F, $Q_{Reflux} = Q_{Dist} = 0$

$$Q_{cond} = 1.995 * 10^6 \text{ Btu/hr}$$

$$\% \text{ Error} = (Q_{in} - Q_{out}) / Q_{in}$$

$$= (1.955 - 1.995) / 1.955 = -2.0\% \text{ (Very Good)}$$

Pressure Drop

The pressure drop is estimated using the vapor and liquid rates and densities provided in the process simulation. For the 2-inch slotted ring, the Generalized Pressure Drop Correlation of Eckert, as modified by Strigle (Figure 13) was used. For the corrugated sheet metal, the manufacturer's pressure drop chart (Figure 15) was used. Results are shown in Tables 7 and 8.

Section	$FP = \dfrac{L}{G}\sqrt{\dfrac{\rho_V}{\rho_L}}$	$C_S\, F_p^{0.5}\, v^{0.05}$	Measured ΔP, in H_2O/ft	Calculated ΔP, in H_2O/ft See Tables 8 and 9
Top Bed	0.0224	0.96	0.14	0.17
Middle Bed	0.0228	0.94	0.18	0.17
Bottom Bed	0.03	Not applicable	0.14	0.13

Table 7. Summary of Measured and Calculated Pressure Drop (Eng Units)

The flow parameter (FP) is often used to describe the hydraulic region of the vapor/liquid or gas/liquid contacting. The flow parameter is determined from the liquid and vapor mass flow ratio and the square root of the vapor density to liquid density ratio.

Location	Top Bed	Middle Bed	Bottom Bed
Test Data			
P, psia	1.97	2.05	2.13
T, °F	173.7	177.3	180.1
Mol. Wt.	106	106	106
Computer Derived Data			
ρ_V, lb/ft^3	0.031	0.032	0.033
ρ_L, lb/ft^3	52.2	52.3	52.4
L, lb/hr-ft^2	858	858	1164
G, lb/hr-ft^2	932	931	999
m_L, cP	0.37	0.38	0.38
GPDC Calculation: from Fig. 13			
FP	0.0224	0.0228	na
F_P (2" rings), ft^{-1}	22	22	na
U_S, ft/s	8.35	8.08	na
Value of Y axis	0.96	0.94	na
ΔP, in H$_2$O/ft	0.17	0.17	na

Structured Packing Calculation from Fig. 15 for Bottom Section

Location	Bottom Bed
U_S, ft/sec	8.4
$C_S = U_S(\rho_V/(\rho_L - \rho_V))^{1/2}$, ft/s	0.20
C_L, gpm/ft^2	2.4
ΔP, in H$_2$O/ft, (Fig.15)	0.13

Table 8. Pressure Drop Estimations Using Data from Figures 13 and 15

Column Capacity

In this example, the column could not be taken to flood because of reboiler and condenser limitations. However, flooding pressure drop for the slotted rings and the structured packing may be assumed to be 1.0 inch water per foot of packing under these conditions. Using the GPDC method for the slotted rings, and the manufacturer's chart for the structured packing, calculations were repeated for estimated conditions that would occur at a pressure drop of 1.0 inch water per foot. The calculated values for the percent of flood was then determined for each bed, and the results are shown in Table 9.

Section	FP	U_S, ft/s	$\sqrt{\dfrac{\rho_V}{\rho_L - \rho_V}}$	C_s, ft/s	$C_{s,\text{flood}}$, ft/s	% of flood
Top	0.0224	8.35	0.0244	0.20	0.38	53
Middle	0.0228	8.08	0.0248	0.20	0.38	53
Bottom	gpm/ft^2 = 2.8	8.33	0.0252	0.21	0.5	42

Table 9. Estimated Percent of Flood for Each Bed Using Data from Figures 13 and 15 (Eng Units)

<u>Column Efficiency</u>

HETP METHOD

Based on the computer simulation, the number of theoretical plates (N_{TP}) required to match the measured separation in each column section is known. From this data, the HETP is calculated as shown in Table 10.

$$HETP = h_T \, / \, N_{TP} \qquad\qquad\qquad \textbf{802.2.1.1}$$

Section	N_{TP}	h_T, ft	*HETP*, ft
Top bed	5	15	3
Middle bed	5	15	3
Bottom bed	5	7.5	1.5

Table 10. Estimated Mass Transfer Efficiency of Each Bed (Eng Units)

The HETP results for the slotted rings compared favorably (36 inches vs 30.5 inches) with the FRI data for a similar system and condition in a tower of the same diameter. The results for the corrugated sheet metal structured packing may be compared with the reported data for the HETP in the cyclohexane/n-heptane system at 4.83 psia and would typically be 16 -18 inches at similar values of C_s.

Data released by FRI for the purposes of this publication are as follows:

Packing:	2-inch metal Pall rings
Distributor:	FRI tubed drip pan
Col. Diameter:	48 inches
Bed Height:	12 feet 0 inches
System:	Cyclohexane / n-Heptane
Pressure:	4.64 psia, top of bed
C_s:	0.134 ft/s
Liq. Rate:	2.3 gpm/ft^2
HETP:	30.5 in.
ΔP:	0.14 in. H_2O/ft

HTU METHOD

As noted by Kister (46), the *HETP/N_{TP}* method is most often used to analyze packed distillation columns, while the HTU/NTU method is most often used for absorbers and strippers. Nevertheless, the HTU/NTU method provides a more fundamental approach with its relationship to the volumetric mass transfer coefficients and mass transfer theory.

Based on values of stripping factor, λ, provided by the process simulation results, the overall height of a vapor phase transfer unit (HTU$_{OV}$) may be determined in the region of the 2 inch slotted rings and the No. 2 structured packing.

Section	λ	HETP, ft
Top Bed – 2 in Rings	1.05	3
Middle Bed – 2 in Rings	1.17	3
Bottom Bed No. 2 Structured Packing	0.97	1.5

Where

$$\lambda = \frac{m \; G_M}{L_M}$$

802.2.1.2

Table 11. Values of Stripping Factors, λ (Eng Units)

For a first approximation, an overall straight equilibrium and operating line are assumed.

$$HETP = \left(\frac{\ln \dfrac{m\,G_M}{L_M}}{\left(\dfrac{m\,G_M}{L_M} \right) - 1} \right) HTU_{OV}$$

802.2.1.3

or

$$HETP = \frac{\left(\ln \{\lambda\} \right)}{\left(\lambda - 1 \right)} HTU_{OV}$$

802.2.1.4

$$HTU_{OV} = \frac{\left(\lambda - 1 \right)}{\ln \{ \lambda \}} HETP$$

802.2.1.5

HTU_{OV}
$= 3.08$ ft (Top Bed)
$= 3.25$ ft (Middle Bed)
$= 1.48$ ft (Bottom Bed)

The number of vapor phase transfer units (NTU_{OV}) can be determined.

$$NTU_{OV} = h_T / HTU_{OV}$$

802.2.1.6

NTU_{OV}
$= 4.9$ (Top Bed)
$= 4.6$ (Middle Bed)
$= 5.1$ (Bottom Bed)

802.2.2 *Packed Column Example calculations with SI Units*

<u>Column Data:</u>

Column Dia (ID)	1.22m	
Packed Sections	Rectifying	Stripping
Packing Type	Slotted Rings	Corrugated Sheet Metal
Packing Size	5 cm	No. 2 (200 m^2/m^3)
No. Beds	2	1
Height each bed	4.57 m	2.29 m

Theoretical Stage Number	Temp, °C	Pressure, mm Hg	Flows, kg/h		Density, kg/m³		p-Xylene Concentration Liquid Wt. Fraction
			Vapor	Liquid	Vapor	Liquid	
Condenser	43.1	99.76	421*	4,210	0.0000	852.3	0.646
1	77.8	99.76	4,631	4,893	0.4886	834.6	0.596
2	78.3	101.3	5,314	4,891	0.4918	834.6	0.548
3	78.7	102.0	5,312	4,889	0.4996	836.2	0.503
4	79.2	102.8	5,310	4,888	0.4998	836.2	0.461
5	79.6	103.5	5,309	4,887	0.5030	836.2	0.424
6	80.0	104.3	5,308	4,886	0.5062	837.8	0.391
7	80.4	105.0	5,307	4,886	0.5110	837.8	0.361
8	80.7	105.8	5,307	4,885	0.5142	837.8	0.336
9	81.0	107.3	5,306	4,885	0.5174	837.8	0.314
10	81.3	108.0	5,306	4,885	0.5223	837.8	0.295
11	81.6	108.8	5,306	6,629	0.5255	837.8	0.279
12	82.0	109.5	5,692	6,630	0.5287	839.4	0.257
13	82.3	110.3	5,693	6,631	0.5335	839.4	0.234
14	82.6	111.0	5,694	6,632	0.5367	839.4	0.210
15	83.0	111.8	5,695	6,633	0.5399	839.4	0.185
Reboiler	83.3	113.3	5,695	938	0.5431	839.4	0.159

*Distillate

Table 12. Process Computer Simulation of O/P Xylene Distillation (SI Units)

Mass Balances

OVERALL MATERIAL BALANCE

% Error = (flow in - flow out) / flow in
= (1362 - 421 – 938) / 1362 = 0.2% (Excellent)

This is within the desired 3%.

COMPONENT MATERIAL BALANCE - P-XYLENE

Feed		= 1362 kg/h
Flow in	= 1362 * 0.31	= 422.2 kg/hr
Flow out, distillate	= 421 * 0.646	= 272.0 kg/hr
Flow out, bottoms	= 938 * 0.159	= 149.1 kg/hr

% Error = (422.2 – 272 – 149.1) / 422.2 = 0.26% (excellent)

This is within the desired 3%.

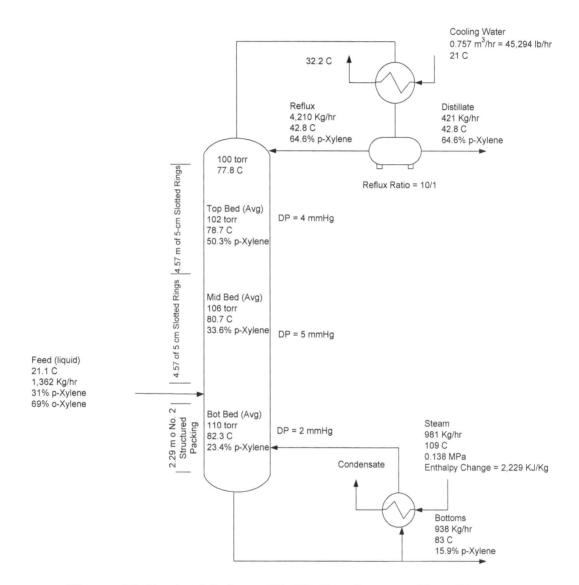

Figure 18. Packed Column Distillation Process Flow Diagram for O/P-Xyelene Distillation (SI Units)

Enthalpy Balance - Datum Condition: liquid at 42.8 °C

OVERALL

$Q_{Reboiler}$	$= 981 * 2229$	$= 2.186 * 10^6$ KJ/hr
Heat Capacity of the feed		$= 1.69$ KJ/Kg-C
Q_{Feed}	$= 1362 * 1.69 * (21.1 - 42.8)$	$= -49,948$ KJ/hr

Heat in $= Q_{Reboiler} + Q_{Feed}$
$= 2.186 * 10^6 - 49,948$
$= 2.136 * 10^6$ KJ/hr

Heat Capacity of Cooling Water $= 4.184$ KJ/Kg-C
$Q_{Cond} = 45,294 * 4.184 * (32.2 - 21.1) = 2.10 * 10^6$ KJ/hr

Heat Capacity of Distillate $= 1.76$ KJ/Kg-C
$Q_{Dist} = 422 * 1.76 * (42.8-42.8) = 0$ KJ/hr

Heat Capacity of Bottoms $= 1.93$ Btu/lb-F
$Q_{Bot} = 938 * 1.93 * (83.3 - 42.8) = 71,319$ KJ/hr

Assume Heat Loss is negligible

Heat out $= Q_{Cond} + Q_{Dist} + Q_{Bot} + Q_{Loss}$
$= 2.1 * 10^6 + 0 + 71,319 + 0$
$= 2.173 * 10^6$ KJ/hr
$= 603.7$ KW

% Error $= (Q_{in} - Q_{out}) / Q_{in}$
$= (593.3 - 603.7) / 593.3 = -1.8\%$ (Very Good)

OVERHEAD LOOP

Vapor Rate $= (R+1) * D = 11 * 421 = 4,631$ Kg/hr
Heat of Vaporization $= 383$ KJ/Kg
Heat in $= Q_{vapor}$
$= 4,642 * 383 + 4,631 * 1.762 * (77.8 - 42.8)$
$= 2.059 * 10^6$ KJ/hr

Heat out $= Q_{Cond} + Q_{Reflux} + Q_{Dist} + Q_{Loss}$

At Reference Temperature of 42.8 C, $Q_{Reflux} = Q_{Dist} = 0$
Q_{Cond} $= 2.1 * 10^6$ KJ/hr
$= 583$ KW

% Error $= (Q_{in} - Q_{out}) / Q_{in}$
$= (572 - 583) / 572 = -1.9\%$ (Very Good)

Column Flows

Location	Top Bed	Middle Bed	Bottom Bed
Test Data			
P, torr	101.9	106	110
T, °C	78.7	80.7	82.3
Mol. Wt.	106	106	106
Computer Derived Data			
ρ_V, kg/m^3	0.496	0.512	0.528
ρ_L, kg/m^3	835	837	839
L, kg/hr-m^2	4,192	4,193	5,688
G, kg/hr-m^2	4,555	4,550	4,882
m_L, cP	0.37	0.38	0.38
GPDC Calculation: from Figure 13 (parameters are in Eng Units)			
FP	0.0224	0.0228	na
F_p (2" rings), ft^{-1}	22	22	na
U_s, ft/s	8.35	8.08	na
Value of Y axis	0.96	0.94	na
ΔP, in H$_2$O/ft (from chart)	0.17	0.17	na
ΔP, mm Hg/m	1.04	1.04	na

Table 13. Pressure Drop Calculation Using Figures 13 and 15 (SI Units)

Structured Packing Calculation from Figure 15 for Bottom Bed:

Location	Bottom Bed
U_S, ft/sec	8.4
$C_S = U_S(\rho_V/(\rho_L - \rho_V))^{1/2}$, ft/s	0.20
C_L, gpm/ft^2	2.4
ΔP, in. H$_2$O/ft (Fig 600.3)	0.13
ΔP, mm Hg/m	0.80

Note: To use Figures 13 and 15, the velocity should have units of ft/s or gpm/ft^2

<u>Column Efficiency</u>

HETP METHOD

Based on the computer simulation, the number of theoretical plates (N_{TP}) required to match the measured separation in each column section is known. From these, the HETP is calculated as shown in Table 14.

$$\text{HETP} = h_T / N_{TP} \qquad\qquad \textbf{802.2.2.1}$$

Section	N_{TP}	h_T, m	HETP, m
Top Bed	5	4.57	0.914
Middle Bed	5	4.57	0.914
Bottom Bed	5	2.29	0.457

Table 14. Estimated Mass Transfer Efficiencies of Each Bed (SI Units)

The HETP results for the slotted rings compared favorably with the FRI data for a similar system and condition in a tower of the same diameter (See note below). The results for the corrugated sheet metal may be compared with the reported data for the HETP in the cyclohexane/n-heptane system at 250 mm Hg would typically be 0.40 – 0.46 m at similar values of C_V and C_L.

Data released by FRI for the purposes of this publication are as follows:

Packing:	2-inch metal Pall rings
Distributor:	FRI tubed drip pan
Col. Diameter:	48 inches (=1.22 m)
Bed Height:	12 feet 0 inches (= 3.66 m)

System:	Cyclohexane / n-Heptane
Pressure:	240 Torr, top of bed
$C_S (= C_V)$::	0.134 ft/s (0.0409 m/s)
Liq. Rate:	2.3 gpm/ft^2 (= 5.62 m^3/m^2-hr)
HETP:	30.5 in. (= 0.775 m)
ΔP:	0.14 in. H$_2$O/ft (= 0.85 mm Hg/m)

HTU METHOD

General suggestions on when to use the HTU Method can be found in 802.2.2.

Based on values of λ provided by a computer simulation, the overall height of a vapor phase transfer unit (HTU$_{OV}$) may be determined in the region of the 2 inch (5 cm) slotted rings and the No. 2 (200 m^2/m^3) structured packing.

where:

$$\lambda = \frac{m\,G_M}{L_M}$$

802.2.2.2

For a first approximation, an overall straight equilibrium and operating line are assumed.

$$HETP = \left(\frac{\ln \dfrac{m\,G_M}{L_M}}{\left(\dfrac{m\,G_M}{L_M} \right) - 1} \right) HTU_{OV}$$

802.2.2.3

or

$$HETP = \frac{\left(\ln \{\lambda\} \right)}{\left(\lambda - 1 \right)} HTU_{OV}$$

802.2.2.4

$$HTU_{OV} = \frac{\left(\lambda - 1 \right)}{\ln\{\lambda\}} HETP$$

802.2.2.5

$$\text{HTU}_{OV} \quad = 0.937 \text{ m (Top Bed)}$$

$$= 0.991 \text{ m (Middle Bed)}$$

$$= 0.45 \text{ m (Bottom Bed)}$$

Section	λ	HETP, m
Top Bed – 2 in Rings	1.05	0.914
Middle Bed – 2 in Rings	1.17	0.914
Bottom Bed – no. 2 Structured Packing	0.97	0.457

Table 15. HTU Method Values of Stripping Factors, λ (SI Units)

The number of vapor phase transfer units can be determined:

$$NTU_{OV} = h_T / HTU_{OV} \qquad \textbf{802.2.2.6}$$

$$\text{NTU}_{OV} \quad = 4.9 \text{ (Top Bed)}$$
$$= 4.6 \text{ (Middle Bed)}$$
$$= 5.1 \text{ (Bottom Bed)}$$

Where:

$$\lambda = \frac{m G_M}{L_M} \qquad \textbf{802.2.2.7}$$

For a first approximation, an overall straight equilibrium and operating line are assumed.

$$HETP = \left(\frac{\ln \dfrac{m G_M}{L_M}}{\left(\dfrac{m G_M}{L_M} \right) - 1} \right) HTU_{OV} \qquad \textbf{802.2.2.8}$$

Or

$$HETP = \frac{\left(\ln\{\lambda\}\right)}{\left(\lambda - 1\right)} HTU_{OV}$$

$$HTU_{OV} = \frac{\left(\lambda - 1\right)}{\ln\{\lambda\}} HETP$$

$$\begin{aligned}
HTU_{OV} \quad &= 0.937 \text{ m (Top Bed)} \\
&= 0.991 \text{ m (Middle Bed)} \\
&= 0.45 \text{ m (Bottom Bed)}
\end{aligned}$$

The number of vapor phase transfer units can be determined:

$$NTU_{OV} = h_T / HTU_{OV}$$

$$\begin{aligned}
NTU_{OV} \quad &= 4.9 \quad \text{(Top Bed)} \\
&= 4.6 \quad \text{(Middle Bed)} \\
&= 5.1 \quad \text{(Bottom Bed)}
\end{aligned}$$

802.3 Trayed Column

The data used in this example are for illustration purposes only and should not be used for design.

The primary separation in the column is methanol and water using a sieve tray distillation column. All of the trace components were ignored to simplify the calculations. Section 14 of *Perry's Chemical Engineers Handbook* [1] is used to estimate percent of flooding and tray pressure drop.

802.3.1 Trayed Column Example Calculations with English Units

Sieve Tray Column Data:

30 Cross-flow Sieve Trays, Feed Tray	= 14
Number of Rectifying Trays	= 13
Number of Stripping Trays	= 17
Tray thickness, in.	= 0.15
Sieve hole diameter, in.	= 0.188
Column diameter (ID), in.	= 66.0
Column area, ft^2	= 23.76
Active area (A_A), ft^2	= 20.48
Downcomer area, ft^2	= 1.64
Hole area (A_h), ft^2	= 3.6

Column diameter (ID), in.	= 66.0
Column area, ft²	= 23.76
Active area (A_A), ft²	= 20.48
Downcomer area, ft²	= 1.64
Hole area (A_h), ft²	= 3.6
Flow path length (FPL), in.	= 50.0
Weir length (L_w), in.	= 43.1
Outlet weir height, in.	= 2.0
Tray spacing, in.	= 24.0

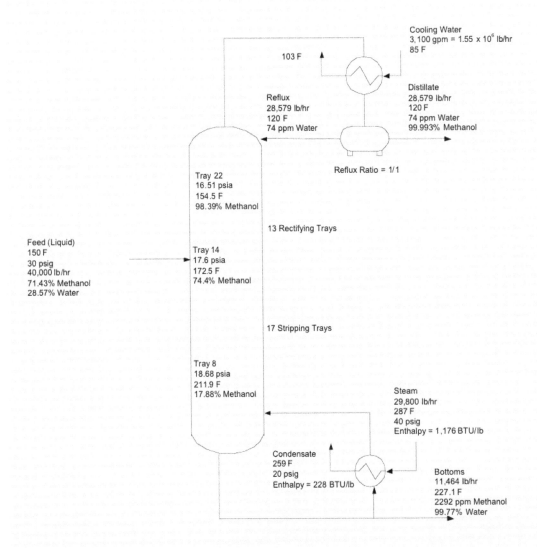

Figure 19. Trayed Column Distillation Process Flow Diagram for Methanol/Water Distillation (Eng Units)

Mass Balances:

OVERALL

$$\% \ error = \frac{flow(in) - flow(out)}{flow(in)} * 100$$

802.3.1.1

$$\% \ error = \frac{40,000 - (28,579 + 11,464)}{40,000} * 100$$

802.3.1.2

% Error = -0.11% (very good)

COMPONENT BALANCES:

Methanol:

$$\% \ error = \frac{flow(in) * X_{feed} - (Dist \ flow * X_{Dist} + Bot \ flow * X_{Bot})}{flow(in) * X_{Feed}} * 100$$

802.3.1.3

$$\% \ error = \frac{40,000 * 0.7143 - (28,579 * 0.99993 + 11,464 * 0.002292)}{40,000 * 0.7143} * 100$$

802.3.1.4

% Error = -0.11% (excellent)

Water:

$$\% \ error = \frac{40,000 * 0.2857 - (28,579 * 0.000074 + 11,464 * 0.99771)}{40,000 * 0.2857} * 100$$

802.3.1.5

% Error = -0.10% (good)

The component balances are excellent.

<u>Enthalpy Balance</u>

OVERALL BALANCE:

Datum condition is liquid at 120°F

Heat in = $Q_{Reboiler} + Q_{Feed}$

$Q_{Reboiler} = 29,047 * (1,176 - 228.0) = 27.54 * 10^6$ BTU/hr

Heat Capacity of Methanol = 0.6 BTU/lb-F

$Q_{Feed} = 0.7143 * 40,000 * 0.60 * (158-120) +$
$0.2857 * 40,000 * 1.0 * (158-120) = 1.0856 * 10^6$ Btu/hr

Heat In = $28.62 * 10^6$ BTU/hr

Heat Out = $Q_{Cond} + Q_{Dist} + Q_{Bot} + Q_{Loss}$ *(Neglect)*

$Q_{Cond} = 1.55 * 10^6 * 1.0 * (103 - 85) = 27.9 * 10^6$ BTU/hr

$Q_{Dist} = 28,600 * 0.69 * (120 - 120) = 0$

$Q_{Bot} = 11,464 * 1.0 * (226 - 120) = 1.22 * 10^6$ BTU/hr

$Q_{Loss} = 0$

Heat out = $29.11 * 10^6$ BTU/hr

$$\% \ Error = \frac{Q_{in} - Q_{out}}{Q_{in}} * 100 \qquad \textbf{802.3.1.6}$$

$$\% \ Error = \frac{28.62 * 10^6 - 29.11 * 10^6}{28.62 * 10^6} * 100 \qquad \textbf{802.3.1.7}$$

% Error = -1.7% (very good)

OVERHEAD LOOP

Datum is liquid at 120 °F

Heat in= $Q_{vapor} = (R+1) * D * (H_{Vap}) + (R+1) * D * Cp * (T_{Vap} - T_{Dist})$

Heat of vaporization = 483.6 BTU/lb

Heat capacity = 0.60 BTU/lb-F

$= 2.0 * 28,579 * 483.6 + 2.0 * 28,579 * 0.6 * (150.4 - 120)$

$$= 28.68 * 10^6 \text{ Btu/h}$$

$$\text{Heat out} = Q_{\text{Cond}} + Q_{\text{Dist}} + Q_{\text{Reflux}}$$
$$= 27.9 * 10^6 + 0 + 0$$
$$= 27.9 * 10^6 \text{ Btu/hr}$$

$$\% \text{ Error} = \frac{Q_{in} - Q_{out}}{Q_{in}} * 100$$

802.3.1.8

$$\% \text{ Error} = \frac{28.68 * 10^6 - 27.9 * 10^6}{28.68 * 10^6} * 100$$

802.3.1.9

% Error = 2.7% (very good)

The overall and local enthalpy balances are very good.

Column Capacity and Flows

The column flooded at rates slightly above those given in the following computer simulation results. These rates are the last stable data and will be used as the basis for a computer simulation of the column. Feed and distillate flow rates, along with the reflux ratio, will be specified. The number of theoretical stages above and below the feed was varied to match the observed concentrations at top and bottom. Table 16 summarizes the computer calculation, showing a column with 20 theoretical stages most closely matching plant data.

The points where column flow rates are highest are equilibrium stages 10 (liquid) and 19 (vapor). For each of these points, a calculation should be carried out to determine the one closest to the flood point. This calculation shows stage 19 is closest to the flood point and is, therefore, likely to flood first. For simplicity, the flooding calculation is shown only for theoretical stage 19.

Percent flood was calculated using Section 14 of *Perry's Chemical Engineers Handbook* and flows for theoretical stage 19.

Theoretical Stage Number	Temp. F	Pressure, psia	Vapor lb/hr	Liquid lb/hr	Vapor Density lb/ft³	Liquid Density lb/ft³	Methanol Liquid wt. fraction
Condenser	120.0		28,579*	28,579	-	49.19	99.9926
20	150.4	15.43	57,158	29,732	0.0761	49.10	99.9699
19	151.0	15.65	58,306	29,712	0.0771	49.10	99.9226
18	151.7	15.86	58.291	29,677	0.0780	49.11	99.8248
17	152.5	16.08	58,256	29,605	0.0790	49.13	99.6236
16	153.4	16.30	58,184	29,460	0.0799	49.16	99.2135
15	154.5	16.51	58,039	29,172	0.0807	49.24	98.3918
14	156.1	16.73	57,752	28,633	0.0812	49.38	96.7970
13	158.5	16.95	57,212	27,705	0.0815	49.66	93.8780
12	162.2	17.16	56,284	26,315	0.0811	50.11	89.0504
11	167.0	17.38	54,894	24,601	0.0801	50.76	82.2276
10	172.5	17.60	53,180	64,858	0.0787	51.53	74.3946
9	175.2	17.81	53,401	63,409	0.0784	51.86	72.0211
8	180.5	18.03	51,592	59,419	0.0767	52.61	63.5627
7	189.5	18.25	47,961	53,970	0.0724	53.99	50.1816
6	201.3	18.46	42,510	48,408	0.0653	55.84	32.9421
5	211.9	18.68	36,947	44,505	0.0578	57.52	17.8821
4	218.9	18.9	33,013	42,409	0.0525	58.6	8.4506
3	222.9	19.11	30,918	41,443	0.0497	59.17	3.6873
2	225.0	19.33	29,980	41,033	0.0487	59.38	1.5376
1	226.2	19.55	29,571	40,872	0.0485	59.48	0.6159
Reboiler	227.1	19.76	29,047	11,464	0.0487	59.52	0.2292

**Distillate Flow

Table 16. Computer Simulation of Methanol/Water Distillation (Eng Units)

$$Flow\ Parameter\ (FP) = \frac{L}{G}\sqrt{\frac{\rho_V}{\rho_L}}$$

802.3.1.10

$$FP = \frac{29,732}{58,306}\sqrt{\frac{0.0771}{49.1}} = 0.0202$$

802.3.1.11

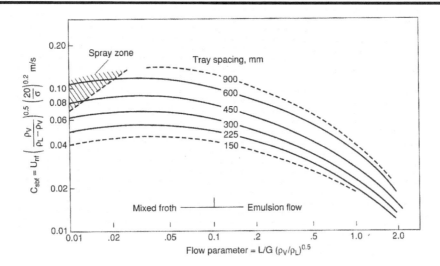

Figure 20. Fair's Entrainment Flooding Correlation for Columns with Crossflow Trays [164]

Using Figure 14.31 of Section 14 in *Perry's Handbook* [1], at a tray spacing of 24 inches, $C_{sbf} = 0.12$ m/s = 0.394 ft/s

$$U_{nf} = 0.394 \sqrt{\frac{(\rho_L - \rho_v)}{\rho_v}} \left[\frac{\sigma}{20}\right]^{0.2}$$

802.3.1.12

$$U_{nf} = 0.394 \sqrt{\frac{(49.1 - 0.0771)}{0.0771}} \left[\frac{20}{20}\right]^{0.2}$$

802.3.1.13

$$U_{nf} = 9.93 \text{ ft/s}$$

802.3.1.14

$$V_{tray19} = \frac{58,306}{23.76 * 0.0771 * 3600} = 8.84 \; ft/s$$

802.3.1.15

$$\% \text{ of flood } = \frac{V_{tray19}}{U_{nf}} *100 = \frac{8.84}{9.93} *100 = 89\%$$

802.3.1.16

Downcomer Flow Limitation

Based on Tray 10 (maximum liquid rate = 64,858 lb/hr)

$$V_{dow} = \frac{W_L}{A_{dow} * \rho_L * 3600}$$

802.3.1.17

$$V_{dow} = \frac{64,858}{1.64 * 51.53 * 3600} = 0.213 \ ft/s$$

802.3.1.18

Using data from Table 14-7 of Section 14 in *Perry's Chemical Engineers Handbook* [1] and based on light hydrocarbon, with 24-inch tray spacing.

$$V_{dow, max} = 0.55 \text{ ft/s}$$

802.3.1.19

$$Downcomer \ Load = \frac{0.213}{0.55} *100 = 38.7\%$$

802.3.1.20

These calculations indicate that the column is near jet, or entrainment flooding, in the top section of the column.

Pressure Drop (h_{total})

For comparison, tray pressure drop should be calculated at several points. The following calculations were done using the pressure drop calculation method described in *Perry's Chemical Engineers Handbook* [1]. We will need to convert to SI units for this method.

$$h_{total} = h_{dry} + h_L^{'}$$

802.3.1.21

At Theoretical Stage 19

W_G (Vapor Mass Flow) = 58,306 lb/hr = 26,470 Kg/hr
W_L (Liquid Mass Flow) = 29,732 lb/hr = 13,498 Kg/hr
ρ_L = 49.1 lb/ft3 = 786.1 kg/m^3
ρ_v = 0.0771 lb/ft^3 = 1.23 kg/m^3

Volumetric Flow rate of Vapor

$$Q_v = \frac{W_G}{\rho_v} = \frac{58,306}{0.0771} = 756,239 \ ft^3/hr = 210 \ ft^3\!/\!{}_s \qquad \text{802.3.1.22}$$

Ah= 3.6 ft2 = 0.3345 m2

$$V_h = \frac{Q_v}{A_h} \qquad \text{802.3.1.23}$$

$$V_h = \frac{210}{3.6} = 58.3 \ ft/s = 17.7 \ m/s \qquad \text{802.3.1.24}$$

$$\frac{A_h}{A_A} = \frac{3.6}{20.48} = 0.176 \qquad \text{802.3.1.25}$$

$$\frac{Tray \ Thickness}{Hole \ Diameter} = \frac{0.15}{0.188} = 0.8 \qquad \text{802.3.1.26}$$

From Figure 14-35 in *Perry's Chemical Engineers Handbook* [1], the discharge coefficient C_v may be determined:

$C_v = 0.82$

For sieve trays and hole velocity in m/s,

$$K = \frac{50}{C_v^2} \qquad \text{802.3.1.27}$$

$$K = \frac{50}{(0.82)^2} = 74.36$$
$$\text{802.3.1.28}$$

$$h_{dry} = KV_h^2 \frac{\rho_v}{\rho_L}$$
$$\text{802.3.1.29}$$

$$h_{dry} = 74.36 * (17.7)^2 * \left(\frac{1.23}{786}\right) = 36.5 \ mmLiquid$$

802.3.1.30

From Equation 14-101 in *Perry's Chemical Engineers Handbook* [1]

$$h_L' = AF * h_{DS}$$

802.3.1.31

where:

$$h_{DS} = h_w + h_{ow} + 0.5 \ h_D'$$

802.3.1.32

h_{ow} = height of crest over weir, mm

$$h_{ow} = 664 \left(\frac{Q_L}{L_w}\right)^{2/3}$$

802.3.1.33

Q_L = Liquid Flow, m³/s

L_w = Weir Length, m

h_D' = Hydraulic Gradient, which was assumed negligible in the example

The aeration factor, *AF*, may be calculated from equation 14-107 of *Perry's Chemical Engineers Handbook* [1]. Units are in metric.

$$AF = 0.0825 \ln \left\{\frac{Q_L}{L_w}\right\} - 0.269 \ln \left\{F_h\right\} + 1.679$$

802.3.1.34

L_w = 43.1 in. = 1.095 m

V_h = 58.3 ft/s = 17.7 m/s

h_w = 50.8 mm

$$Q_L = \frac{W_L}{\rho_L} = \frac{29,732}{49.1} = 605.5 \ ft^3/hr = 0.0048 \ m^3/s$$

802.3.1.35

$$F_h = V_h \sqrt{\rho_v} = (17.7) \sqrt{1.23} = 19.6 \; m/s \sqrt{kg/m^3}$$

802.3.1.36

$$AF = 0.0825 \ln \left\{ \frac{0.0048}{1.095} \right\} - 0.269 \ln \{19.6\} + 1.679$$

802.3.1.37

$$AF = 0.431$$

$$h_{total} = h_{dry} + AF \left[664 \left(\frac{Q_L}{L_w} \right)^{2/3} + h_w \right]$$

802.3.1.38

$$h_{total} = 36.5 + 0.431* \left[664 \left(\frac{0.0048}{1.095} \right)^{2/3} + 50.8 \right]$$

802.3.1.39

$$h_{total} = 66.0 \; mm \; Liquid = 2.59 \; in \; Liquid$$

802.3.1.40

$$h_{total} = 2.59 \left(\frac{\rho_L}{\rho_{water}} \right) = 2.59 \left(\frac{49.1}{62.4} \right) = 2.04 \; in \; H_2O$$

802.3.1.41

Additional Pressure Drop Points:

Theoretical Stage	h_{total} (in. of water)
19	2.04
10	2.22
1	1.73
Average	2.00

Column Efficiency

The overall column efficiency can be calculated from the previous computer calculation.

OVERALL COLUMN

$$Efficiency = \frac{\text{\# of theo. stages}}{\text{\# of actual trays}} * 100$$

802.3.1.42

$$Efficiency = \frac{10}{13} * 100 = 76.9\%$$

802.3.1.43

Based on the process simulation, the individual section efficiencies may also be calculated:

TOP (RECTIFYING) SECTION

$$Efficiency = \frac{20}{30} * 100 = 66.7\%$$

802.3.1.44

BOTTOM (STRIPPING) SECTION

$$Efficiency = \frac{10}{17} * 100 = 58.8\%$$

802.3.1.45

Checking efficiency using the O'Connell graphical correlation (Figure 21, based on [166]).

Average Relative volatility of the methanol/water system (α) in stripping section ~ 4

Average relative volatility of the methanol/water system (α) in the rectifying section ~ 1.2

Average liquid viscosity (μ_L) ~ 0.3 cP

Therefore (μ_L) x (α) = 0.3 * 4 = 1.2 (Stripping Section)

= 0.3 * 1.2 = 0.36 (Rectifying Section)

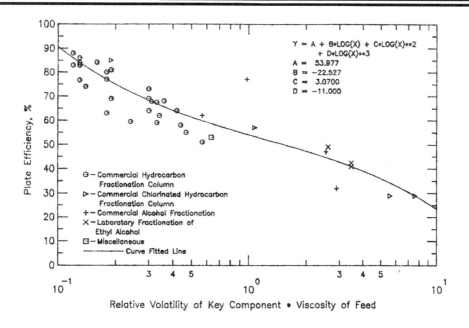

Figure 21. Tray Efficiency using O'Connell Plot ([165], reprinted from [166])

From Figure 21 above, the predicted tray efficiency is approximately 54% in the stripping section, and 66% in the rectifying section, both of which match reasonably well with the test data.

802.3.2 *Trayed Column Example Calculation with SI Units*

Sieve Tray Column Data:

30 cross-flow sieve trays, feed tray	= 14
Number of rectifying trays	= 13
Number of stripping trays	= 17
Tray thickness, cm	= 0.38
Sieve hole diameter, cm	= 0.48
Column diameter (ID), m	= 1.68
Column area, m^2	= 2.207
Active area (A_A), m^2	= 1.903
Downcomer area, m^2	= 0.152
Hole area (A_h), m^2	= 0.3345
Flow path length (FPL), m	= 1.27
Weir length (L_w), m	= 1.095
Outlet weir height, cm	= 5.08
Tray spacing, m	= 0.61

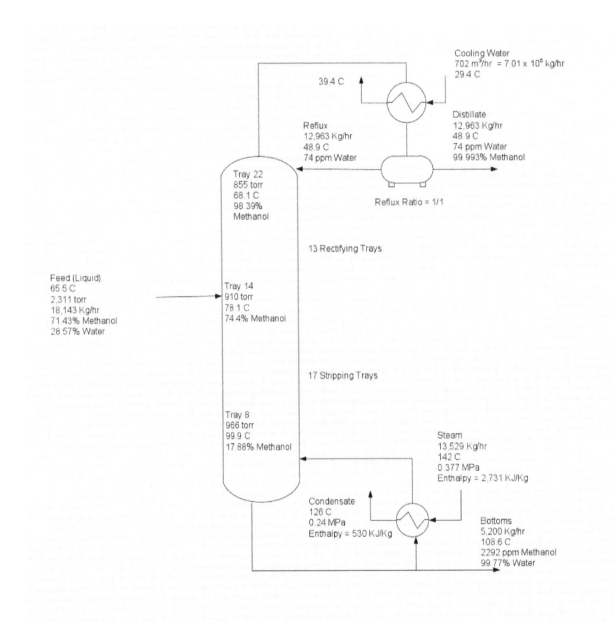

Cooling Water
702 m³/hr = 7.01 x 10⁶ kg/hr
29.4 C

39.4 C

Reflux
12,963 Kg/hr
48.9 C
74 ppm Water

Distillate
12,963 Kg/hr
48.9 C
74 ppm Water
99.993% Methanol

Tray 22
855 torr
68.1 C
98.39%
Methanol

Reflux Ratio = 1/1

13 Rectifying Trays

Feed (Liquid)
65.5 C
2,311 torr
18,143 Kg/hr
71.43% Methanol
28.57% Water

Tray 14
910 torr
78.1 C
74.4% Methanol

17 Stripping Trays

Tray 8
966 torr
99.9 C
17.88% Methanol

Steam
13,529 Kg/hr
142 C
0.377 MPa
Enthalpy = 2,731 KJ/Kg

Condensate
126 C
0.24 MPa
Enthalpy = 530 KJ/Kg

Bottoms
5,200 Kg/hr
108.6 C
2292 ppm Methanol
99.77% Water

Figure 22. Trayed Column Distillation Process Flow Diagram for Methanol/Water Distillation (SI Units)

Material Balance:

OVERALL

$$\% \ error = \frac{flow\,(in) - flow\,(out)}{flow\,(in)} * 100 \qquad \textbf{802.3.2.1}$$

$$\% \ error = \frac{18,143 - (12,963 + 5,200)}{18,143} * 100 \qquad \textbf{802.3.2.2}$$

% Error = -0.11% (very good)

COMPONENT BALANCES:

Methanol:

$$\% \ error = \frac{flow\,(in) * X_f - (Dist\ flow * X_D + Bot\ flow * X_B)}{flow\,(in) * X_f} * 100 \qquad \textbf{802.3.2.3}$$

$$\% \ error = \frac{18,143 * 0.7143 - (12,963 * 0.99993 + 5,200 * 0.0023)}{18,143 * 0.7143} * 100 \qquad \textbf{802.3.2.4}$$

% Error = -0.12% (excellent)

Water:

$$\% \ error = \frac{18,143 * 0.2857 - (12,963 * 0.000074 + 5,200 * 0.9977)}{18,143 * 0.2857} * 100 \qquad \textbf{802.3.2.5}$$

% Error = -0.11% (excellent)

Energy Balances

OVERALL BALANCE:

Datum Condition is Liquid at 48.9°C

$\qquad$ Heat in = $Q_{Reboiler}$ + Q_{Feed}

$\qquad\qquad Q_{Reboiler}$ = $13,339 * (2,731 - 530) = 29.36 * 10^6$ KJ/hr

$\qquad\qquad$ Heat Capacity of Methanol = 2.5 KJ/Kg-C

$$\text{Heat Capacity of Water} = 4.18 \text{ KJ/Kg-C}$$

$$Q_{Feed} = 0.7143 * 18{,}143 * 2.5 * (65.5 - 48.9) +$$
$$0.2857 * 18{,}143 * 4.18 * (65.5 - 48.9)$$
$$= 0.897 * 10^6 \text{ KJ/hr}$$

$$\text{Heat in} = 30.26 * 10^6 \text{ KJ/hr} = 8{,}405 \text{ KW}$$

$$\text{Heat out} = Q_{Cond} + Q_{Dist} + Q_{Bot} + Q_{Loss}(Neglect)$$
$$Q_{Cond} = 7.02 * 10^5 * 4.184 * (39.4 - 29.4) = 29.33 * 10^6 \text{ KJ/hr}$$
$$Q_{Dist} = 12{,}963 * 2.5 * (48.9 - 48.9) = 0$$
$$Q_{Bot} = 5{,}200 * 4.184 * (108.39 - 48.9) = 1.294 * 10^6 \text{ KJ/hr}$$
$$Q_{Loss} = 0$$

$$\text{Heat out} = 30.62 * 10^6 \text{ KJ/hr} = 8{,}506 \text{ KW}$$

$$\% \ Error = \frac{Q_{in} - Q_{out}}{Q_{in}} * 100$$

802.3.2.6

$$\% \ Error = \frac{8405 - 8506}{8405} * 100$$

802.3.2.7

$$\% \ Error = -1.2\% \ (\text{very good})$$

OVERHEAD LOOP

$$\text{Heat in} = Q_{vapor} = (R+1)*D*(H_{Vap}) + (R+1)*D*Cp\,(T_{Vap} - T_{Dist})$$
$$\text{Heat of vaporization} = 1{,}123 \text{ KJ/Kg}$$
$$\text{Heat capacity} = 2.5 \text{ KJ/Kg-C}$$
$$= 2.0 * 12{,}963 * 1{,}123 + 2.0 * 12{,}963 * 2.5 * (65.77 - 48.9)$$
$$= 30.21 * 10^6 \text{ KJ/hr} = 8{,}391 \text{ KW}$$

$$\text{Heat out} = Q_{Cond} + Q_{Dist} + Q_{Reflux}$$
$$= 29.33 * 10^6 + 0 + 0$$
$$= 29.33 * 10^6 \text{ KJ/hr} = 8{,}147 \text{ KW}$$

$$\% \ Error = \frac{Q_{in} - Q_{out}}{Q_{in}} * 100$$

802.3.2.8

$$\% \ Error = \frac{8391 - 8147}{8391} * 100$$

802.3.2.9

% Error = 2.9% (very good)

Both the overall and local enthalpy balances are very good.

<u>Column Capacity and Flows</u>

The column flooded at rates slightly above those given in Table 17. These rates are the last stable data and will be used as the basis for a computer simulation of the column. As with the previously documented results in English units, the feed and distillate flow rates and the reflux ratio was specified, and the number of theoretical stages above and below the feed was varied to match the observed concentrations at top and bottom. A column with 20 theoretical stages most closely matched plant data.

As in the earlier calculations, stages 10 (liquid) and 19 (vapor) has the highest column loads, and a flood calculation for each of these points was performed. This calculation showed that stage 19 is closer to the flood point and is therefore likely to flood first. Flooding calculation is shown only for theoretical stage 19.

Column Data:

30 cross-flow one pass sieve trays, feed tray	= 14
Number of rectifying trays	= 13
Number of stripping trays	= 17
Tray thickness	= 3.8 mm
Sieve hole diameter	= 4.8 mm
Column diameter, ID	= 1.676 m.
Column area	= 2.207 m^2
Active area	= 1.902 m^2
Downcomer area	= 0.152 m^2
Hole area	= 0.335 m^2
Flow path length	= 1.27 m
Weir length	= 1.09 m
Outlet weir height	= 50 mm
Tray spacing	= 0.610 m

Theoretical Stage Number	Temp. C	Pressure, mm Hg	Vapor kg/hr	Liquid kg/hr	Vapor Density kg/m^3	Liquid Density kg/m^3	Methanol Liquid wt. fraction
Condenser	48.9		12,963*	12,963	–	787.9	99.9926
20	65.77	798.1	25,926	11,486	1.219	786.5	99.9699
19	66.11	809.3	26,447	13,477	1.235	786.5	99.9226
18	66.50	820.6	26,440	13,461	1.249	786.6	99.8248
17	66.94	831.8	26,424	13,429	1.265	787.0	99.6236
16	67.44	843.1	26,392	13,363	1.280	787.4	99.2135
15	68.06	853.6	26,326	13,232	1.293	788.7	98.3918
14	68.94	865.6	26,196	12,988	1.301	791.0	96.7970
13	70.28	876.8	25,951	12,567	1.305	795.0	93.8780
12	72.33	887.3	25,530	11,936	1.299	802.7	89.0504
11	75.00	898.6	24,899	11,159	1.283	811.1	82.2276
10	78.00	909.8	24,122	29,419	1.261	825.4	74.3946
9	79.55	921.1	24,222	28,598	1.256	830.7	72.0211
8	82.50	932.3	23,402	26,952	1.229	842.7	63.5627
7	87.50	943.6	21,755	24,480	1.160	864.8	50.1816
6	94.06	954.8	19,282	21,957	1.046	894.4	32.9421
5	99.94	966.1	16,759	20,187	0.926	921.4	17.8821
4	103.83	977.3	14,988	19,236	0.841	938.7	8.4506
3	106.06	988.6	14,037	18,798	0.796	947.3	3.6873
2	107.22	999.8	13,599	18,613	0.78	951.1	1.5376
1	107.89	1011.1	13,413	18,539	0.777	952.8	0.6159
Reboiler	108.39	1021.6	13,339	5,200	0.78	953.4	0.2292

*Distillate Flow

Table 17. Computer Simulation of the Methanol/Water Distillation (SI Units)

As shown in Section 802.3.1, the calculated percent flood for theoretical stage 19 is 89%

PRESSURE DROP

The tray pressure drop should be calculated at several points for comparison. The correlations used are given in *Perry's Chemical Engineers' Handbook* [1]. The calculation procedure is the same as shown in Section 802.3.1. The results of the calculation are summarized below.

Additional pressure drop points:

Theoretical Stage	P (mm Hg)
19	3.81
10	4.13
1	3.23
Average	3.73

COLUMN EFFICIENCY

Column efficiency can be calculated from the previous computer calculation.

Overall Column

$$Efficiency = \frac{\# \ of \ theo \ stages}{\# \ of \ actual \ trays} * 100$$

802.3.2.10

$$Efficiency = \frac{20}{30} * 100 = 66.7\%$$

802.3.2.11

The individual section efficiencies based on the process simulation may also be calculated:

Top Section

$$Efficiency = \frac{10}{13} * 100 = 76.9\%$$

802.3.2.12

Bottom Section

$$Efficiency = \frac{10}{17} * 100 = 58.8\%$$

802.3.2.13

From Figure 21, the predicted tray efficiency is approximately 54% in the stripping (bottom) section and 66% in the rectifying (top) section which match reasonably well with the test data.

803.0 References

1. Perry, R.H. and Don Green, *Perry's Chemical Engineers' Handbook,* 8th ed. New York: Mc-Graw-Hill, 2008.

2. McCabe, W.L. and J.C. Smith, *Unit Operations of Chemical Engineering*, New York: McGraw-Hill, 1976.

3. Hengstebeck, R.J., *Petrol. Eng.*, 29, C-6, 1957.

4. Bolles, W.L. and J.R. Fair, *IChemE Symposium Series,* 56, 3.3/35, 1979.

5. Shaw, R.J., J.A. Sykes and R.W. Ormsby, *Chem. Eng.*, 87(16) Aug. 11, 1980, 126.

6. Kister, H.Z., *Distillation Operation*, New York: McGraw-Hill, 1990.

7. McLaren, D.B., and J.E. Upchurch, *Chem. Eng.*, 77(12), June 1, 1970, 139-152.

8. Sands, R.R., "Column Instrumentation Basics: Understanding Accuracy and Sources of Error." Presented at the Distillation Topical Conference, AIChE Spring Meeting, Houston, TX, April 2007.

9. Zuiderweg, F.J. (Editor), *Recommended Test Mixtures for Distillation Columns*, European Federation of Chemical Engineering, Inst. of Chem. Engrs., London, 1969.

10. Herington, R.F. (Editor), "Recommended Reference Materials of Realization of Physico Chemical Properties, Section: Testing Distillation Columns," *Pure and Applied Chem*, 51 1979, 2421-2499.

11. "Sensible Heat Transfer in Shell and Tube-Type Equipment" *AIChE Standard Testing Procedure for Heat Exchangers:* Section I.

12. ASME Research Committee on Fluid Meters Report, *Fluid Meters-Their Theory and Applications*, 6th Ed., 1971.

13. Head, V.P., *Trans. ASME*, 76, 1954, 851.

14. Cai, T.J., A. Shariat and M.R. Resetarits, Distillation Symposium of the 2009 Spring AIChE Meeting, Tampa, Florida.

15. Bernard, A., W. DeVilliers and D.R. Summers, *Hydrocarbon Processing*, April 2009, 61-69.

16. Perry, R.H. *Chemical Engineers' Handbook*, 5th Edition, 1973, 5-6.

17. Silvey, F.C. and G.J. Keller, *Chem. Eng. Progress*, 62 (1), 1966, 62-68.

18. Kastenek, F., and G. Standard, *Separation Sci.*, 2(4), 1967, 439.

19. Lockett, M.J., and I.S. Ahmed, *Chem. Eng. Res. Des.*, 61(2), 1983, 110.

20. Kister, H.Z., *Chem. Eng.*, 88(3), Feb. 9, 1981,107-109.

21. Taylor, R. and R. Krishna, *Multicomponent Mass Transfer*, Hoboken, NJ: John Wiley and Sons, Inc, 1993.

22. Bolles, W.L., "How to Conduct Distillation Column Flood Capacity Tests," personal communication.

23. Kelley, R.E., T.W. Pickel and G.W. Wilson, *Petrol. Refiner*, 34(1), 1955, 110; 34(2), 1955, 159.

24. Lipowicz, M.A., *Chem. Eng.*, 91(1), 1984, 66-68.

25. Xu, S.X., and L. Pless, "Distillation Tower Flooding – More Complex Than You Think," *Chem. Eng. Progress*, 98(6), June 2002, 60-67.

26. Branan, C., *The Fractionator Analysis Pocket Handbook*, Houston, TX: Gulf Publishing, 1978.

27. Gibson, G.J., *Chem. Eng.*, 94(7), May 11, 1987, 75.

28. Lieberman, N.P., *Troubleshooting Process Operations*, 2nd Ed., Tulsa, OK: Penwell Publishing, 1985.

29. Yarborough, L., L.E. Petty and R.H. Wilson, *Proc 59th Annual Convention of the Gas Processors Assoc*, Houston, Texas, March 17-19, 1980, 86-89.

30. Robinson, E.S., and E.R. Gilliland, *Elements of Fractional Distillation*, 4th Ed., New York: McGraw-Hill, 1950, 476-478.

31. Summers, D.R., "Evaluating and Documenting Tower Performance," *Chem. Eng. Progress*, 106(2), Feb. 2010, 38-45.

32. Gelus, E., S. Marple, Jr., and E. Manning, Jr., *Chem. Eng. Progress*, 45(10), Oct. 1949, 602-608.

33. Taylor, R., R. Krishna and H. Kooijman, "Real-World Modeling of Distillation," *Chem. Eng. Progress*, 99(7), July 2003, 28-39.

34. Distillation Subcommittee of the Research Committee, *Bubble Tray Design Manual*, New York: AIChE, 1958.

35. Bolles, W.L., "Optimum Bubble Cap Tray Design," Fritz W. Glitsch & Sons, Inc., Bulletin No. 156-2, Dallas, TX.

36. Bolles, W.L., "Tray Hydraulics-Bubble Cap Trays," Ch. 14 in Smith, B.D., *Design of Equilibrium Stage Processes*, New York: McGraw-Hill, 1963, 474.

37. Kister, H.Z., and J.R. Haas, "Predict Entrainment Flooding on Sieve and Valve Trays", *Chem. Eng. Progress*, 86(9), Sept. 1990, 63-69.

38. Fair, J.R., "Tray Hydraulics-Perforated Trays," Ch. 15 in Smith, B.D., *Design of Equilibrium Stage Processes*, New York: McGraw-Hill, 1963, 539-569.

39. Bolles, W.L., *Chem. Eng. Progress*, 72(9), Sept. 1976, 43-49.

40. Klein, G.F., *Chem. Eng.*, 89(9), May 3, 1982, 81-85.

41. Thorngren, J. T., *Ind. Eng. Chem. Proc. Des. Dev.*, 11(3), 1972, 428-429.

42. Kister, H. Z. and J. R. Haas, "Entrainment From Sieve Trays in the Froth Regime," *Ind. Eng. Chem. Res.*, 27(12), December 1988, 2331-2341.

43. Glitsch, Inc., "Ballast Tray Design Manual," Bulletin No. 4900, 3rd Ed., Dallas, TX.

44. Koch Engineering Co., Inc., "Flexitray Design Manual," Bulletin 960-T. Wichita, KS.

45. Nutter Engineering Co., "Float Valve Design Manual" Tulsa, OK.

46. Kister, H.Z., *Distillation Design*, New York: McGraw-Hill, Inc., 1992.

47. Strigle, R.F., Jr., *Random Packings and Packed Towers*, Houston, TX: Gulf Publishing Co., 1987.

48. Baker, T. and J.S. Stockhardt, *Ind. Eng. Chem.*, 22(4), 1930, 376.

49. Gerster, J.A., T. Mizushina, T.N. Marks and A.W. Catanach, *AIChE Journal*, 1(4), 1955, 536.

50. Taylor, D.L., P. Davis, and E.D. Holland, *AIChE Journal*, 10(6), 1964, 864; 11(4), 1965, 678.

51. Sherwood, TK., Pigford, R.L. and C.R. Wilke, *Mass Transfer*, New York: McGraw-Hill,1975, 427-428.

52. Peters, W.A ., *Ind. Eng. Chem.*, 14(6), June 1922, 476-479.

53. McCabe, W.L. and E.W. Thiele, *Ind. Eng. Chem.*, 17 (6), June 1925, 605-611.

54. Wisniak, J., et al., "Binary Distillation," *AIChE Modular Instructions*, Series B, Vol. 1, 1980.

55. King, C.J., *Separation Processes,* 2nd Edition, New York: McGraw-Hill Inc., 1980, 140-501.

56. Thompson, R.E., "Multicomponent Distillation," *AIChE Modular Instruction*, Series B, Vol. 2, 1981.

57. Holland, C.D., *Fundamentals of Multicomponent Distillation*, New York: McGraw-Hill, Inc., 1981.

58. Holland, C.D., *Fundamentals and Modeling of Separation Processes, Absorption, Distillation, Evaporation, and Extraction*, Upper Saddle River, NJ: Prentice-Hall, Inc., 1975.

59. Van Winkle, M., *Distillation*, New York: McGraw-Hill, Inc., 1967.

60. Watkins, R.N., *Petroleum Refinery Distillation, 2nd Edition*, Houston, TX: Gulf Publishing Co., 1978.

61. Chilton, T.H. and A.P. Colburn, *Ind. Eng. Chem.*, 27(3), March 1935, 255-260.

62. Koshy, T.D. and F. Rukovena, "Effect of Concentration and Reflux Ratios on Distillation Tower Design," AIChE Annual Meeting, Los Angeles, Dec. 1982.

63. Perry, R.H, and C.H. Chilton, *Chemical Engineers' Handbook*, 6th Edition, New York: McGraw-Hill, 1984, 13-96.

64. Lewis, W.K., and W.G. Whitman, *Ind. Eng. Chem.*, 16(12), December 1924, 1215-1220.

65. Fair, J. et al., "Gas Absorption and Gas-Liquid System Design", Section14-8, in Perry, R.H., *Chemical Engineers' Handbook*, 7th Edition, N.Y. 1997.

66. Billingsley, D.S., A. Chirachavala, *AIChE Journal*, 27(6), 1981, 968-974.

67. Krishnamurthy, R. and R. Taylor, *AIChE Journal*, 31(3), 1985, 448.

68. Krishnamurthy, R. and R. Taylor, *AIChE Journal,* 31(3), 1985, 456.

69. Sivasubramanian, M.S. and J.F. Boston, "Rate- based Separation Modeling Techniques," AIChE Annual Meeting, Paper, No. 82d, November 27, 1988.

70. Sivasubrilmanian, M.S. and J.F. Boston, "A Rate-based Approach for Modeling Multicomponent Separation Processes," CACHI 89 European Symposium on Computer Application in the Chemical Industry, Erlangen, Federal Republic of Germany, April 23-26, 1989.

71. Seader, J.D., "The Rate-based Approach for Modeling Staged Separation-A New Era," *Chem. Eng. Progress,* 85(10), October 1989, 41.

72. Ludwig. E.E., *Applied Process Design for Chemical and Petrochemical Plants*, Vol. 2, 2nd Ed., Houston, TX: Gulf Publishing Co., 1979.

73. Perry, R.H., *Chemical Engineers' Handbook*, 5th Edition, New York: McGraw-Hill, Inc., 1973, 18.34-18.48.

74. Leva, M., *Tower Packing and Packed Tower Design*, 2nd Ed., United States Stoneware, Co., 1953, 91.

75. McAdams, W.H., et al., *Chemical Engineering Progress*, 44 (4), 1948, 241-252.

76. Mickley, H.S., *Chem. Eng. Progress*, 41(12), 1945, 738-745.

77. Coulson, J.M. and J.F. Richardson, Chapter 10, *Chem. Eng.*, Vol. 1, Oxford, UK: Pergamon Press, 1978.

78. Eckert, J.S., *Chem. Eng. Progress*, 71(8), 1975, 60-68.

79. McNulty, K.J. and C.L. Hsieh, "Hydraulic Performance and Efficiency of Koch Flexipac Structured Packings, AIChE Annual Meeting, Los Angeles, Dec. 1982.

80. Fair, J.R., *Chem. Eng.*, 79(13), 1972, 91-100.

81. Fair, J.R., "Design of Direct-Contact Gas Coolers," *Petro/Chem Engineer*, 33(9), August 1961, 57-64.

82. Stockar, U.V. and C.R. Wilke, *Ind. Eng. Chem. Fundamental*, 16(1), 1977, 94-103.

83. Billet, R., M. Schultes, "Prediction of Mass Transfer Columns with Dumped and Arranged Packings," *Trans. IChemE*, 77(6), September 1999, 498-504.

84. Fair, J.R. and J.L. Bravo, "Distillation Columns Containing Structured Packing," *Chem. Eng. Progress,* 86(1), Jan 1990, 19-29.

85. Custer, R.S., *Chem. Eng. Progress*, 61 (9), 1965, 86.

86. Berg, C., and I.J. James, Jr., *Chem. Eng. Progress*, 44 (4), 1948, 307.

87. Shah, G.C., *Chem. Eng.*, 85 (17), July 31, 1978, 70.

88. Martin, H.W., *Chem. Eng. Progress*, 60 (10), 1964, 50.

89. Resetarits, M., K. Chuang and N. Yang, "Determination of Number of Theoretical Plates in Distillation Column with commercial Simulation Software," Annual AIChE Meeting, Los Angeles, CA, November 1997.

90. Bravo, J.L., J.A. Rocha and J.R. Fair, "Pressure Drop in Structured Packings," *Hydrocarbon Processing*, March 1986, 45-49; *Distillation and Absorption*, 1987, A233-244.

91. Ergun, S., "Fluid Flow Through Packed Columns," *Chem. Eng. Progress*, 48(2), Feb. 1952, 89.

92. Buchanan, J.E., "Pressure Gradient and Liquid Holdup in Irrigated Packed Towers," *I & E C Fund*, 8(3), 1969, 502-511

93. Bemer, G.G. and G.A.J. Kalis, "A New Method to Predict Holdup and Pressure Drop in Packed Columns," *Trans I Chem E*, 56, 1978, 200-204.

94. Takahashi, T., Y.A. Akagi and K. Ueyama, "A New Correlation for Pressure Drop of Packed Column," *J Chem Engr. Japan*, 12 (5), 1979, 341-346.

95. Fair, J.R. and J.L. Bravo, 'Prediction of Mass Transfer Efficiencies and Pressure Drop for Structured Tower Packing in Vapor/Liquid Service," *I. Chem. E. Distillation and Absorption 1987*, A183.

96. Perry, R.H., *Perry's Chemical Engineer's Handbook*, 7th Ed., New York: McGraw-Hill, Inc., 1998.

97. Hausch, D.E., *Chem. Eng. Progress*, 60 (10), 1964, 55.

98. Buckley, P.S., R.K. Cox and D.L. Rollins, *Chem. Eng. Progress*, 71(6), 1975, 83.

99. Troyan, J.E., "Troubleshooting New Equipment," *Chem. Eng.*, 68(6), March 20, 1961, 147-150.

100. Formisano, F.E., "Method Quickly Troubleshoots Packed-Column Problems," *Chem. Engr.*, 76(11), Nov. 3, 1969, 108-110.

101. Drew, J.W., "Distillation Column Startup," *Chem. Eng.*, 90(11), Nov. 14, 1983, 221-226.

102. Kister, H.Z. and T.C. Hower, "Unusual Operating Histories of Gas Processing and Olefins Plant Columns," AIChE Annual Meeting, Miami Beach, FL., Nov. 2-7, 1986.

103. Kunesh, J.G., "Recent Developments in Packed Columns," *Canadian J of Chem. Engr.*, 65(12), Dec. 1987, 907-913.

104. Billet, R., "Recent Investigations of Metal Pall Rings," *Chem. Eng. Progress*, 63 (9), 1967, 53-65.

105. Billet, R., "Optimization and Comparison of Mass Transfer Columns," *I. Chem. E. Symposium Series* No. 32, London, 1969.

106. Eckert, J.S., "Selecting the Proper Distillation Column Packing," *Chem. Eng. Progress*, 66 (3), 1970, 39-44.

107. Nygren, P.G. and G.K.S. Connolly, "Selecting Vacuum Fractionation Equipment," *Chem. Eng. Progress*, 67 (3), March 1971, 49-58.

108. Charpentier, J.C., "Recent Progress in Two-Phase Gas-Liquid Mass Transfer in Packed Beds," *Chemical Engineering Journal*, 11(3), 1976, 161-181.

109. Harrison, M.E. and J.J. France, "Troubleshooting Distillation Columns, Part 1, Technique and Tools," *Chem. Eng.* 96 (3), 1989; "Part 2, Packed Columns," 96 (4), 1989, 121-128; "Part 4, Column Auxiliaries," 96 (6), 1989, 130-137.

110. Kister, H.Z., *Distillation Trouble Shooting*, Hoboken, NJ, John Wiley & Sons, Inc., 2006.

111. Eagle, R.S., *Chem. Eng. Progress*, 60(10), 1964, 69.

112. Glausser, W.E., *Chem. Eng. Progress*, 60(10), 1964, 67.

113. Snow, A.I. and W.S. Dickinson, *Chem. Eng. Progress*, 60(10), 1964, 64.

114. Jamison, R.H., *Chem. Eng. Progress*, 65(3), 1969, 46.

115. Formisono, F.A., *Chem. Eng.*, 76(24), Nov. 3, 1969, 108.

116. Kister, H.Z., *Chem. Eng.*, 87(26), Dec. 29, 1980, 55.

117. Kister, H.Z., *Chem. Eng.*, 87(15), July 28, 1980, 79.

118. Kister, H.Z., *Chem. Eng.*, 87(10), May 19, 1980, 138.

119. Kister, H.Z., *Chem. Eng.*, 87(18), Sept. 8, 1980, 119.

120. Thorogood, R.M., *I. Chem. E. Symp. Ser.*, 61, 1981, 95.

121. Fair, J.R., and R.L. Matthews, *Petrol Refiner*, 37(4), 1958, 153.

122. Sakata, M., *Chem. Eng. Progress*, 62 (11), 1966, 98.

123. Kister, H.Z., *Hydroc. Proc.*, 58(2), 1979, 89.

124. Sherwood, T.K. and F.A.L. Holloway, "Performance of Packed Towers-Liquid Film Data for Several Packings," *Trans Am. Inst. Chem. Engrs.*, 36 (21), 1940, 39.

125. Yoshida, F. and T. Koyanagi, "Liquid Phase Mass Transfer rates and Effective Interfacial Area in Packed Absorption Columns," *Ind. Eng Chem.*, 50 (3), March, 1958, 365-374.

126. Strigle, R.F. and K.E. Porter, *I. Chem. E. Symp. Ser.* No. 32, 1979, 3119.

127. Shulman, H.L. et al., *AIChE Journal*, 1(2), 1955, 247-264.

128. Onda, K., H. Takeuchi, and Y. Okumoto, "Mass Transfer Coefficients between Gas and Liquid Phases in Packed Columns," *J. Chem. Engr. Japan*, 1 (56), 1968.

129. Hughmark, G.A., "Mass Transfer and Flooding in Wetted-Wall and Packed Columns," *I & E C Fund.*, 19(4), 1980, 385-389.

130. Zuiderweg, F.J, *Chem Eng Sci*, 37, 1982, 1441.

131. Bolles, W.L. and J.R. Fair, "Improved Mass-Transfer Model Enhances Packed Column Design," *Chem. Eng.*, 89 (14), 1982, 109-116.

132. Bravo, J.L. and J.R. Fair, "Generalized Correlations for Mass Transfer in Packed Distillation Columns," *I&E C Process Design and Development*, 21(1), 1982, 162-170.

133. Zuiderweg, F.J., P.J. Hoek and L. Lahm, "The Effect of Liquid Distribution and Redistribution on the Separating Efficiency of Packed Columns," *I. Chem. E. Distillation and Absorption 1987*, A217-223.

134. Kunesh, J.G., L.L Lahm and T. Yanagi, "Controlled Maldistribution Studies on Random Packing at a Commercial Scale," *I. Chem. E. Distillation and Absorption 1987*, A233-244.

135. Krishna, R., "A Film Model Analysis of Nonequimolar Distillation of Multicomponent Mixtures," *Chemical Engineering Sci.*, 32(6), 1977, 1197.

136. Smith, L.W. and R. Taylor, "Film Models for Multicomponent Mass Transfer: A Statistical Comparison," *I&E C Fund*, 22 (1), 1983, 97.

137. Billet, R., *Packed Towers*, in *Processing and Environmental Technology*, VCH Verlagsgesellschaft mbH, Weinheim, FRG,1995.

138. Billet, R., *Distillation Engineering*, Hoboken, NJ: John Wiley & Sons, Inc., 1979.

139. Andersen, A.E. and J.E. Jubin, *Chem. Eng. Progress*, 60 (10), 1964, 60.

140. Loud. G.D. and R.C. Waggoner, *Ind. Eng. Chem. Proc. Des. Dev.*, 17(2), 1978, 149.

141. Kunesh, J.G., L. Lahm and T. Yanagi, "Commercial Scale Experiments Provide Insight on Packed Tower Distributors," *I&EC Research* 26(9), 1987, 1846.

142. Kunesh, J. G., L.L. Lahm and T. Yanagi, "Controlled Maldistribution Studies on Random Packings at a Commercial Scale," International Conference on Distillation and Absorption 1987, *I.Chem.E Symposium Series* No. 104, A233-A244.

143. Fitz, C. W., D.W. King and J.G. Kunesh, "Controlled Liquid Maldistribution Studies on Structured Packing," *Trans IChemE*, 77, Part A, 1999, 482-486.

144. Cai, T. J., G.X. Chen, C.W. Fits and J.G. Kunesh, "Effect of Bed Length and Vapor Maldistribution on Structured Packing Performance," International Conference on Distillation and Absorption, 2002, 2.2-1.

145. Schultes, M., "Influence of Liquid Redistributors on the Mass-Transfer Efficiency of Packed Columns," *Ind. Eng. Chem.*, 2000, 39 (5), 1381-1389.

146. Lockett, M.J. and J.F. Billingham, "The Effect of Maldistribution in Separation in Packed Distillation Columns," International Conference on Distillation and Absorption 2002, 2.2-2.

147. Moore, E. and F. Rukovena, "Liquid and Gas Distribution in Commercial Packed Towers," 36th Canadian Chemical Engineering Conference, Paper 23b, October 5-8, 1986.

148. Perry, D., D.E. Nutter, and A. Hale, "Liquid Distribution for Optimum Packing Performance," *Chem. Eng. Progress,* 86 (1), Jan 1990 30.

149. Bolles, W. L., "Multipass Flow Distribution and Mass Transfer Efficiency for Distillation Plates," *AIChE Journal*, 22 (1), 1976, 153-158.

150. Pilling, M., "Ensure Proper Design and Operation of Multipass Trays," *Chem. Eng. Progress*, June 2005, 22-27.

151. Kister, H.Z., R. Dionne, W.J.Stupin and M. Olsson, M., "Preventing Maldistribution in Multipass Trays," Distillation Topical Conference, AIChE Spring Meeting, Tampa, Florida, April 2009.

152. Summers, D.R., "Three Pass Trays – Friends or Foe?" AIChE Annual Meeting, Paper 21a, Nashville, Tennessee, November 2009.

153. Summers, D.R., "Designing Four-Pass Trays," *Chem. Eng. Progress*, April 2010, 26-31.

154. Kister, H.Z., *Chem. Eng.*, 88(7), Apr. 6, 1981, 97.

155. Bolles, W.L., *Chem. Eng. Progress*, 63(9), 1967, 48.

156. Barker, P. E., *Brit. Chem. Eng.*, 8(5), 1963, 306.

157. Davies, J.A., *Chem. Eng. Progress*, 61(9), 1965, 74.

158. Grohse, E.W., R.F. McCartney, H.J. Hauer, J.A. Gerster and A.P. Colburn, *Chem. Eng. Progress*, 45 (12), 1949, 725.

159. Lockett. M.J., C. T. Lim. and K.E. Porter, *Trans. Instn. Chem. Engrs.*, 51,1973, 61.

160. Lockwood, D.C. and W.E. Glausser, *Petrol. Refiner*, 38(9), 1959, 281.

161. Love, F.S., *Chem. Eng. Progress*, 71(6), 1975, 61.

162. Gagneux, A.L. and A.B. Hiser, *Petrol. Refiner*, 33(3), 1954, 165.

163. Ognisty, T. P. and M. Sakata, "Multicomponent Diffusion: Theory vs. Industrial Data," *Chem. Eng. Progress*, 83, March1987, 60-65.

164. Fair, J.R., "How to Predict Sieve Tray Entrainment and Flooding," *Petro/Chem Engineer,* 33(10), September 1961, 45.

165. O'Connell, H.E., *Trans AIChE*, 42, 1946, 741-755.

166. Fair, J.R., "Distillation and Gas Absorption," Chapter 13, *Chemical Process Handbook-Selection and Design*, 2nd Ed., 2005.

INDEX